M

Fire Loss Control

OCCUPATIONAL SAFETY AND HEALTH

A Series of Reference Books and Textbooks
on Occupational Hazards ● Safety● Health ●
Fire Protection ● Security ● and Industrial Hygiene

Series Editor
ALAN L. KLING
Loss Prevention Consultant
Jamesburg, New Jersey

Additional Volumes in Preparation

Fire Loss Control

A Management Guide
Second Edition
Revised and Expanded

Peter M. Bochnak

Harvard University
Cambridge, Massachusetts

MARCEL DEKKER, INC.   New York • Basel • Hong Kong

Library of Congress Cataloging--in--Publication Data

Bochnak, Peter M.
  Fire loss control: a management guide/ Peter M. Bochnak. -- -- 2nd
ed., rev. and expanded.
    p.   cm. -- -- (Occupational safety and health; 22)
  Includes bibliographical references and index.
  ISBN 0-8247-8413-8
  1. Industrial buildings-- --Fires and fire prevention. I. Title.
II. Series: Occupational safety and health (Marcel Dekker, Inc.);
22.
  TH9445.M4B63   1990
  628.9'22- --dc20   90--23460
                                                          CIP

This book is printed on acid-free paper.

MARCEL DEKKER, INC.
270 Madison Avenue, New York, New York 10016

Current printing (last digit):
10 9 8 7 6 5 4 3 2 1

PRINTED IN THE UNITED STATES OF AMERICA

For their support
during the writing of this book,
I dedicate this edition to
Marilyn, Steve, and Susan

In Memoriam

ROBERT G. PLANER

The tragic and untimely death of the author and creator of the first edition of this text was a great loss to us personally and to the fire protection community.

The first edition stands as proof of the many contributions Robert Planer made to fire protection technology and particularly, as a fire protection engineer and to the teaching of this discipline.

Peter M. Bochnak

Preface to the Second Edition

The objective of this book is to provide the manager, the architect, the plant engineer, the technician, and others with a background in fire protection in as concise a manner as possible and to point out other sources of information available for more in-depth information. Good fire protection practices necessitate the informed concern of all individuals who have management or design responsibilities over facilities. It is earnestly hoped that managers, architects, engineers, students, technologists, and other readers will obtain a realistic assessment of fire protection and property loss control so that all may have a broad, commonly based, view of this area of property conservation.

The book has been completely updated. New statistical data have been provided, new references added, and new methods or standards shown where growth in knowledge or actions by standard-making bodies has occurred since the first edition. All the chapters have been rewritten to reflect current fire protection and fire prevention practices, and there are five new chapters: "Life Safety Elements," "Emergency Organization," "Common Process Hazards," "Water Supplies for Fire Loss Control," and "Education and Training."

Except for temperature conversions, English units of measurement are used in this book. English to metric (SI) unit conversion factors for more commonly used fire protection measurements can be found in the Appendix.

Peter M. Bochnak

Preface to the First Edition

When approached to write this book, my instantaneous reaction was one of acceptance, quickly followed by an inner warning not to react too quickly. Vanity put to route the warning, and with the acceptance of the responsibility came the laborious efforts to put together something which readers hopefully would find valuable.

In my contacts as a property loss control consultant for Johnson & Higgins, working with architects, engineers, and other consultants for clients, or with client personnel themselves, I found over an extensive period of time a surprising lack of knowledge of the need for adequate levels of fire protection and good property loss control practices in general. More often than not, this extended to management personnel involved with projects whose costs were often staggering. Many projects in which Johnson & Higgins was involved were unique, representing key investments which were sometimes the sole production facility for a product line of considerable importance to the profit picture of the client, meriting at the very least a surface knowledge of good fire protection by those responsible for the overall project direction.

This general lack of knowledge was also found in the students whom I was fortunate enough to contact as the coordinator and head instructor in a course in fire protection technology given as a Continuing Engineering Studies course at the New Jersey Institute of Technology.   Many of the participants in the program had over the years assumed loss control responsibilities for firms such as insurance companies and fire protection bureaus before attending this course.

Although numerous works on fire protection are available, I felt that there was a place for a book which would give the manager, architect, engineer, students, and technologist a broader view of the field of property conservation. This book is principally written as a tool for such people and others seeking a general treatment of this topic.

Except for temperature conversions, English units of measurements are used exclusively in this book. Recognizing the growing importance and value of metric (SI) units, the reader will find conversion factors for more commonly used fire protection measurements at the end of the text.

One never writes a book without being indebted to many who have provided encouragement or assistance. Alan L. Kling is basically responsible for this book, which was done at his request and which was made a reality by his continuing assistance and encouragement. Encouragement has also been provided by William T. Dunn, Jr. and Thomas E. Barton, Senior Vice Presidents and Directors, and John D. Crawford, Vice President of Johnson & Higgins, whose support is greatly appreciated. The necessary mechanical work was supplied by Josephine Russo and Mack Martin, who worked tirelessly on the manuscript. My sincere thanks seems very little reward for their efforts.

To my wife, Jane, who has had to put up with a great deal during the working period, I dedicate this book. I would also like to do honor to all those fire protection engineers who labor in the "vineyards," frequently being misunderstood and considered a nuisance, as "cost increasers." It is my sincere hope that they will find the book worthy of them.

Robert G. Planer

Contents

Fire Loss Control

1

Introduction

Today, with the real presence of the Occupational Safety and Health Administration (OSHA) and the employee's "right" to safety and health, coupled with the "age of risk management," captive insurance companies, and large self-retention of risk by major corporate entities, there is a greater need for management to familiarize itself with the general field of fire protection. In the past, there appears to have been a general feeling that the protection of plants and facilities from fire and the maintenance of their integrity could be safely left to insurance underwriter personnel or perhaps to a very small staff or an individual who might have something less than a heavy background in this area. Now, with greater retention of risk by corporations and the tremendous investments involved in many facilities, there is a need for many more people, at the very least, to have more than a superficial knowledge of the world of fire protection.

This need has impacted not only upon plant management, but also upon architects and engineers. Many architectural engineering firms have found it prudent to employ specialists in fire protection on their staffs to handle what should be considered a very important part of the overall plant design. It is important for these technical people to develop some knowledge of fire protection, since many firms of this type are not only providing fire protection concepts, but in many cases are doing the actual design, although they may not possess a high degree of fire protection expertise.

Risk management, with its attempts to bring a degree of sophistication to decisions on handling and preserving corporate assets, must by necessity begin with personnel who have the knowledge of fire protection necessary to evaluate the risks facing a corporation. Who better can perform this task evaluating risks in the property area than an expert in fire prevention? Once decisions are made to retain the risk, transfer the risk in the form of insurance, or handle it in other ways, property loss control still should play a most important part in risk control. Hopefully, this will be recognized by the decision makers in risk management.

Where does the risk manager turn—or, for that matter, the plant manager, technician, architect, or engineer who needs to know something about fire protection? Do busy managers attempt to wade through 12 volumes of the *National Fire Codes* published by the National Fire Protection Association (NFPA)? Do they read from cover to cover the NFPA *Fire Protection Handbook* (1760 pages)? Do they go through nine volumes of the Factory Mutual *Loss Prevention Data Books*? Many publications are available on this very broad subject that, for the most part, are directed at those having more than a casual interest in this field. Many of these publications are voluminous and represent complete works either on the entire scope of fire protection or on specific areas. For the professional in fire protection, such as the fire protection engineer, a library of these publications is indispensable.

However, turning back to the question as to where the busy manager should turn for some background, there appear to be few publications available that can give a relatively short treatment or an overview on this subject. So we now come to the objective of this book, which is to provide the manager, the architect, the plant engineer, the technician, and others with a background in fire protection in as brief and concise a manner as possible and then point out other sources available for more in-depth information.

The seasoned fire protection engineer who will pick up this book may find little new, although at the same time he or she may find certain things that, while generally understood throughout the field, may not have been put in print previously or covered in such a manner. One such area is the section on "site selection" related to fire protection and somewhat closely allied areas, such as seismic and flood plain exposures, which are very important inputs into site determinations for major facilities. Another is the interrelations of the OSHA fire protection regulations and the overall fire loss control program.

This publication is, therefore, specifically directed to management-oriented personnel with an involvement in loss prevention; to fire protection technicians who are not engineering oriented, but who nevertheless must have a relatively strong background in fire protection; to the architects and engineers involved with overall plant design; and, last but not least, to students of fire protection technology.

Those who may have the impression that fire protection is a very narrow field have no concept of its broad scope, the areas on which it impacts, and the extent of knowledge that a professional in this field must have. This includes not only expertise in a specific field such as hydraulics, but also in the design and installation of adequate protection systems for the hazards that may be involved.

SOURCES OF INFORMATION

Before exploring the specifics of managing fire losses, one should have some knowledge of the organizations that play an important role in influencing the field of fire protection and, more importantly, what information can be obtained from these sources. In the following sections, these organizations will be discussed briefly, including their backgrounds, available information and its importance, and where the material can be obtained. This discussion will include the NFPA, Underwriters Laboratories Inc. (UL), fire protection testing laboratories, insurance organizations, and other organizations, all of which have a strong influence on fire protection. This is not a complete listing, but does indicate the variety of organizations that have an interest in fire protection and loss control.

Fire Protection Organizations

National Fire Protection Association, Batterymarch Park, Quincy, MA 02169

Today, more than any other organization, the NFPA influences the field of fire protection not only nationally, but throughout the world, by its consensus standard-making activity; by its investigation of significant fire losses; and by its large number of publications, which range from the release of standards to training aids for the fire services, the publication of the *Fire Protection Handbook*, and the disseminating of fire protection material. The NFPA was established in 1896 as a scientific and educational membership organization concerned with the causes, prevention, and control of fire.

The NFPA publishes annually the *National Fire Codes*, in a 12-volume compilation set as well as in individual pamphlet form. These volumes include numerous consensus standards that have been adopted over the years and updated. These include standards on the installation of extinguishing systems, including automatic sprinklers; on flammable liquids; on spray finishing and dip tanks; on combustible solids, gases, and dusts; on electrical installations; on construction; and numerous other standards.

Many of these standards have been adopted by various governmental agencies and, consequently, are law, including NFPA standards that have been referenced or adopted by OSHA in its regulations. Proposed new standards or revisions are contained in *Technical Committee Reports* and *Technical Committee Documentation*.

Information on the latest techniques in fire prevention, protection, and suppression and information on significant fires and fire-related developments are contained in the NFPA periodicals *Fire Journal, Fire News,* and *Fire Technology.* NFPA publications also include the *Fire Protection Handbook*, occupancy fire safety studies and reports, various publications on fire services training, and a number of handbooks based on the more popular NFPA standards.

Membership in the NFPA is open to individuals, corporations, firms, institutions, municipal departments, and others. In addition to individual membership, organization membership in the NFPA is open to international, national, and regional societies and associations interested in the protection of life and property against loss by fire.

Society of Fire Protection Engineers, 60 Batterymarch St., Boston, MA 02210

The Society of Fire Protection Engineers (SFPE) was established in 1950 and is the professional society for fire protection engineers. The SFPE serves as a clearinghouse for fire protection engineering state-of-the-art advances and information and publishes the *SFPE Bulletin* and a series of *Technology Reports.* The SFPE and NFPA jointly publish the *SFPE Handbook of Fire Protection Engineering*, which covers the theory, research results, and calculation methods of fire protection engineering.

Membership in the SFPE is open to people with engineering or physical science qualifications and experience in the fire protection field.

Insurance Organizations

Insurance organizations and companies, particularly those referred to as the HPR (highly protected risk, or highly preferred risk) carriers, have traditionally been in the forefront of fire protection and fire loss control. They provide engineering and inspection services to their clients, develop standards, and provide research in fire protection and publications on this subject, largely in the field of industrial loss prevention.

Factory Mutual System, 1151 Boston-Providence Turnpike, Norwood, MA 02062

The Factory Mutual (FM) System was founded in 1835 and is composed of three large mutual property insurance companies—Allendale Insurance, Arkwright, and Protection Mutual Insurance—and two system associates—Factory Mutual International and Factory Mutual Engineering and Research. The FM System insists upon good loss control practices as a requirement for issuance of coverage. The system was developed on the basis of preventive care and protection equipment. Today, this organization conducts basic and applied research, develops standards, issues approvals on various materials and equipment, and provides insurance at cost through loss prevention inspections of plants, research, and consultation services to its insureds. Its engineering group does this through evaluation of the hazards involved; the protection systems; and management's property conservation methods, or the "human element" of fire protection.

FM publishes the *Loss Prevention Data* books in a nine-volume set that is available to companies insured by the FM System, risk management consultants, public fire departments, government bodies, educational institutions, and various trade associations. This service covers a wide range of fire protection, including construction, sprinkler protection, water supplies, extinguishing equipment, industrial hazards, etc. Numerous other loss prevention materials are published, including the *Handbook of Property Conservation*. FM also publishes a periodical, *The Record*, which contains articles on property conservation engineering and management.

Industrial Risk Insurers, 85 Woodland St., Hartford, CT 06102

The Industrial Risk Insurers (IRI) was formed in 1975 as a result of the merger of the Factory Insurance Association (FIA) and the Oil Insurance Association (OIA). FIA, one of the forerunners of IRI, was founded in 1890 to conserve from loss the plants and production of American industries. Today, IRI is an international oraganization insuring properties in over 70 countries.

IRI, along with other HPR carriers, generally requires that its accounts meet the following criteria: (1) be sprinklered industrial properties with hazards adequately protected; (2) have management that demonstrates its willingness and determination to reduce the probability of loss; (3) be of a class where specialized underwriting, inspection, engineering, and loss prevention services will be of value to the policyholder; and (4) generate sufficient premium flow to justify the costs of these services.

IRI operates a Loss Prevention Training Center in its main office primarily for the training of its own engineers, but also offers intensive courses in the proper use of fire protection devices for plant personnel of IRI policyholders and for representatives of member companies, agents, and brokers. IRI publishes a periodical, *The Sentinel*, that covers information of current interest in the field of industrial loss prevention. IRI also publishes a manual entitled *Overview: A Management Program for Loss Prevention and Control* covering practices they recommend for certain industries, problem areas, or specific hazards.

Insurance Services Office, 160 Water St., New York, NY 10038

The Insurance Services Office (ISO) was formed in 1971 with the merger of a number of insurance industry service and rating organizations nationwide. ISO is an association of approximately 1300 affiliated insurance companies engaged in casualty and property insurance. It functions as an insurance rating organization, as an insurance service organization, and as a statistical agent. ISO publishes the "Commercial Fire Rating Schedule" and the "Fire Suppression Rating Sched-

an association of approximately 1300 affiliated insurance companies engaged in casualty and property insurance. It functions as an insurance rating organization, as an insurance service organization, and as a statistical agent. ISO publishes the "Commercial Fire Rating Schedule" and the "Fire Suppression Rating Schedule." In addition, ISO handles the municipal fire protection grading of cities and towns.

American Insurance Association, 85 John St., New York, NY 10038

The American Insurance Association (AIA) is a trade association serving a large number of companies in the property and casualty insurance fields. The AIA was created in 1964 with the merger of the National Board of Fire Underwriters (founded 1866), the Association of Casualty and Surety Companies (founded 1926), and the former American Insurance Association (founded 1953). AIA provides a range of services, including legislative services, engineering and safety services, a property insurance loss register, and research and review of claim and loss adjustment functions. The AIA Engineering and Safety Service offers guidance to its subscriber companies in total loss control and publishes a number of Bulletins on fire protection, safety, and casualty topics. AIA's "Special Interest Bulletins," which contain important fire protection information, are available to all fire protection interests.

Other Insurance Groups

In addition to FM and IRI, two other major influences in the HPR field are Improved Risk Mutuals and Kemper Insurance Companies.

Improved Risk Mutuals, located in White Plains, New York, was founded in 1921 and consists of 18 member companies that reinsure large industrial, commercial, and institutional properties. It publishes engineering bulletins on numerous topics related to fire protection. Kemper Insurance Companies, located in Long Grove, Illinois, publishes the *Kemper Report*, a periodical that includes items of interest on fire protection and related topics.

Major insurance brokers are an excellent source of assistance in property loss control through their large staffs of specialists employed in this area. These specialists or consultants are assigned to specific clients to provide consultation as part of the broker's commission. While brokers do not conduct fire research, they do possess a broad knowledge of information sources and can use the experience of many engineering disciplines and specialties to solve client problems. A helpful pamphlet, "The Broker's Role in Risk Management," may be obtained from Johnson and Higgins, 95 Wall Street, New York, NY. This publication describes how large insurance brokers operate and what can be expected from the services they provide.

Testing Laboratories

Of utmost importance to users of various materials for construction, to those installing fire protection, or to those involved with equipment interfacing with hazardous substances is that the materials exhibit acceptable flammability characteristics, have satisfactory reliability, have certain characteristics necessary for specific fire protection applications, and are safe when used with hazardous substances such as flammable liquids, gases, etc. This assurance comes from using products that have been tested and accepted by a reliable organization. Such organizations traditionally have been known as nationally recognized testing laboratories (NRTL). This term is seen in a number of local, state, and federal regulations and codes, where authorities will accept products and assemblies that are listed or approved by a NRTL.

In 1988, OSHA established regulations for testing laboratories (29CFR 1910.7) to be recognized as an NRTL by OSHA when the term "nationally recognized testing laboratory" is used in OSHA standards. Among the laboratories that have received the recognition are Underwriters Laboratories Inc. and Factory Mutual Research Corporation, two organizations with historical preeminence in the fire testing field.

Underwriters Laboratories Inc., 333 Pfingsten Road, Northbrook, IL 60062

Underwriters Laboratories Inc. founded in 1894, is an independent organization testing for public safety. UL maintains and operates laboratories for the examination and testing of devices, systems, and materials to determine their relation to life safety, fire, casualty hazards, electrical hazards, and crime prevention. UL has organized engineering councils in the areas of casualty, burglary protection, electrical, fire, and marine that review reports on devices prior to their release. These councils represent a broad spectrum of membership having expertise in their specific areas.

UL publishes annual directories of materials and devices it has tested. Following is a list of the directories available:

Building Materials Directory
Fire Protection Equipment Directory
Fire Resistance Directory
Electrical Appliance and Utilization Equipment Directory
Electrical Construction Materials Directory
Hazardous Location Equipment Directory
Marine Products Directory
Classified Products Directory
Accident, Automotive, and Burglary Protection Equipment Directory
Gas and Oil Equipment Directory

Of the above directories, persons involved with plant construction will be specifically interested in the *Building Materials Directory* and *Fire Resistance Directory*, which are described in Chapter 8, Fire Safety in Design and Construction. Of particular importance in the installation of fire protection equipment, including extinguishers, automatic sprinklers, or other protective systems, is the *Fire Protection Equipment Directory*.

Factory Mutual Research Corporation, 1151 Boston-Providence Turnpike, Norwood, MA 02062

The Factory Mutual Research Corporation is the research and testing arm of the Factory Mutual System. Its function is to conduct research and development in the area of property loss control. This research activity is available not only to the Factory Mutual System, but also to other groups, such as government agencies, trade associations, and businesses, through contracts. A major effort is maintained in applied research, including studies, surveys, operations research, experimentation, and full-scale testing to evaluate hazards and protection schemes.

As a result of its extensive testing program, FM Research Corporation publishes annually an *Approval Guide* that is available to FM insureds at no charge and to others at a nominal charge. This approval guide, similar to UL directories, lists protective equipment, combustion safeguards, building materials and construction, etc., that have been tested and either approved or accepted. Perhaps a distinction should be made between FM's "approval" and its "acceptance." "Approval" refers to a product that has been tested according to FM standards and found suitable for general application, while "acceptance" refers to a specific installation or arrangement of equipment or material. Devices listed in the *Approval Guide* are "approved." Installations incorporating these devices that are found to be satisfactory following a review of plans and inspection of completed work are "accepted." Approved equipment bears the Factory Mutual System approval marks.

Government Agencies

Government agencies are becoming more and more involved in fire prevention and protection. These agencies are discussed in Chapter 14, Regulatory Contacts.

2

Characteristics and Behavior of Fire

Those involved in the prevention, control, or suppression of fire should have an appreciation of the characteristics and behavior of fire, as well as knowledge of how fire may intensify or spread and how it may be controlled and ultimately extinguished. The chemistry and physics of fire are extremely complex. However, an understanding of the fundamental technical aspects of fire, based on the accumulated experience of those engaged in fire prevention and control, can help in appreciating the problems of fire loss management and suggesting practical solutions.

The chemistry of fire involves the combining of a combustible material with oxygen, called *oxidation*. The rusting of iron and steel is an example of a very slow form of oxidation. Fire or combustion is the rapid oxidation of a material accompanied by heat and usually light. The oxidation reaction involved is *exothermic*, which means that energy is being released in the form of heat, as opposed to an *endothermic* reaction, in which energy is absorbed. To have this rapid oxidation, called fire or combustion, certain elements must be present:

A combustible material, also called fuel.
An oxidizing agent; in most cases, oxygen in our atmosphere.
A specific temperature; fuel must be raised to this temperature so that the heat input will be sufficient to sustain combustion.

Traditionally, the fire triangle was used to depict these requirements for combustion: fuel, oxygen, and temperature. As shown in Figure 2.1, one side repre-

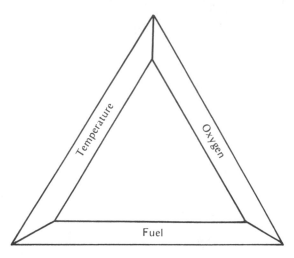

Figure 2.1   Fire triangle showing the basic requirements for combustion.

sents the fuel component, one side the oxygen component, and the third side the temperature component. By removing any of these elements, a fire can be extinguished. Using water to extinguish a fire through hose lines, sprinklers, or water spray cools the burning material, thereby removing the temperature side of the triangle. Extinguishing agents such as foam or carbon dioxide prevent the oxygen from reaching the fire and thus remove the oxygen side of the triangle. Removing the fuel side, such as by shutting off the flow of a flammable gas or liquid, is the third way to extinguish a fire.

Later, it was discovered that dry chemical and halogenated agents extinguish fires by inhibiting the chain reaction of the combustion process. This added a new element, uninhibited chain reaction, to the original triangle, which was redesignated as the fire tetrahedron to explain the combustion process.

This concept was refined to show that the process occurs in two modes: flaming and surface (1) (see Fig. 2.2). The flaming mode has high burning rates with heat being fed back to the burning material to sustain combustion. Equilibrium is reached when the heat energy generated balances with the energy lost to the environment. If the heat energy generated is greater, the fire will grow. As more energy is lost to the environment, the fire will decrease. The flaming mode can be explained by the fire tetrahedron (see Fig. 2.3).

The surface mode (which includes glowing and deep-seated glowing embers) consists of radiative feedback to sustain ignition. This type of fire does not rely on the chain reaction and can be explained by the fire triangle (see Fig. 2.3). The two modes are not mutually exclusive; they can each exist in combination, as they do,

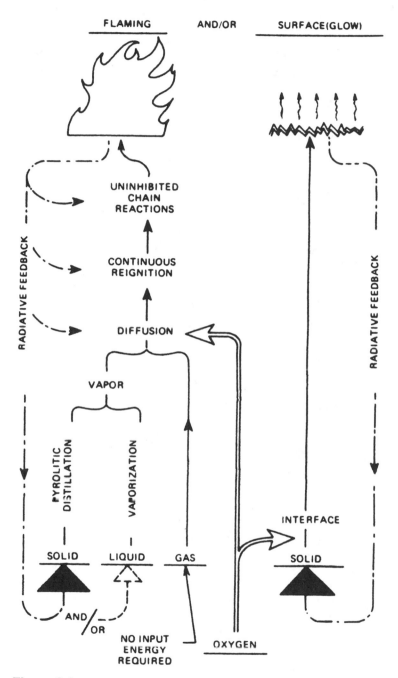

Figure 2.2 Basic fire system modes. (Reprinted with permission from the *Fire Protection Handbook,* 15th Edition, Copyright 1981, National Fire Production Association, Quincy, MA 02269.)

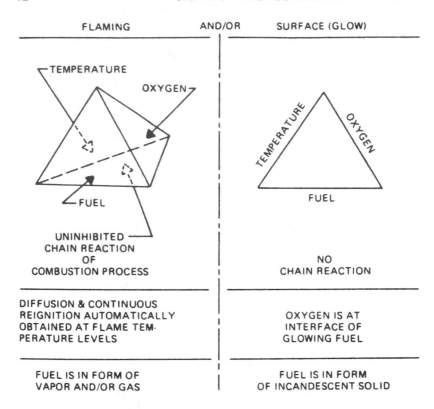

FLAMING   AND/OR   SURFACE (GLOW)

Figure 2.3 Basic fire system mode requirements. (Reprinted with permission from the *Fire Protection Handbook*, 15th Edition, Copyright 1981, National Fire Protection Association, Quincy, MA 02269.)

for example, with wood and coal. Flammable liquids and gases burn in the flaming mode, and combustible metals, carbon, and oxidizable nonmetals such as phosphorous and sulphur burn only in the surface combustion mode (1).

Although fire normally involves rapid oxidation of a fuel by oxygen in the air, combustible metals such as magnesium, aluminum, and cadium can "burn" in an atmosphere of pure nitrogen. Some fuels, such as cellulose nitrate, contain their own oxygen. Nitromethane, hydrogen peroxide, and ozone can directly decompose, emitting light and heat.

When determining what extinguishing agent should be used on a fire, it is important to know the classes of fires (2).

Class A fires occur in ordinary combustible materials such as wood, cloth, and paper. Extinguishing this type of fire requires the cooling and quenching action of water, although a multipurpose dry chemical is also effective because it provides a rapid knock-down of flame and forms a fire-retardant coating that prevents reflash.

Class B fires occur in flammable and combustible liquids. Extinguishing this type of fire requires the oxygen-excluding (smothering) or combustion-inhibiting effects provided by dry chemical, foam, vaporizing liquids, and carbon dioxide.

Class C fires occur in energized electrical equipment. This type of fire requires the use of nonconducting extinguishing agents, such as dry chemical, carbon dioxide, and vaporizing liquids.

Class D fires occur in combustible metals such as magnesium, titanium, zirconium, and sodium. Specialized techniques, extinguishing agents, and extinguishing equipment have been developed to control and extinguish this type of fire. Water and normal extinguishing agents should not be used on this type of fire.

With this in mind, it is important to know some of the chemistry and physics of fire. For flammable liquids, we start with an explanation of vapor pressure and boiling point. As liquid molecules leave the surface of an open container, they form a vapor. Since the container is open, the liquid evaporates. With a closed container, the vapor is limited to the space above the liquid. At the point of equilibrium (equal amount of vapor leaving and entering the liquid), the pressure exerted is the vapor pressure, measured in psia (pounds per square inch atmospheric) or kPa (kiloPascals). As the temperature of a liquid increases, the vapor pressure increases. The boiling point of a liquid is the temperature at which vapor pressure equals atmospheric pressure. The percentage of vapor is directly proportional to the relationship between the vapor pressure and the total pressure of the vapor-air mixture above the liquid.

Another important characteristic is the *flash point* of a liquid. This is the lowest temperature at which the vapor pressure of the liquid is sufficient to produce a flammable mixture. There are various ASTM test standards and devices available to measure flash points (3). The type of test used depends on the flash-point range and the type of liquid. The flash point can vary with the pressure and oxygen content of the atmosphere.

The *fire point* of a liquid is the lowest temperature at which a vapor-air mixture will continue to burn after it is ignited. This is generally a few degrees above the flash point.

Knowing vapor pressure at the flash point, one can calculate the lower flammable limit (LFL). LFL = $100V/P$, where V is the vapor pressure (psia) and P is the ambient pressure (psia). For acetone at 0°F (−18°C) and a vapor pressure (*V*) of 0.38 psia at an ambient pressure (*P*) of 14.7 psia, the LFL would equal 2.6% (percent vapor by volume):

$$\text{LFL} = \frac{100(0.38)}{14.7} = 2.6\%$$

The LFL is the minimum concentration of vapor in air below which the propagation of flame will not occur. The upper flammable limit (UFL) is the maximum concentration above which propagation of flame will not occur. These limits are also known as the lower explosive limit (LEL) and upper explosive limit (UEL).

The upper flash point is the maximum temperature above which the vapor-air concentration is too high to propagate flame. The relationships among flash point, vapor pressure, temperature, and concentration are shown in Figure 2.4. As can be seen in this figure, as the temperature increases, the flammable range widens.

While the flash point is the lowest temperature at which a flammable vapor-air mixture can exist, the *ignition temperature* is the minimum temperature to which a substance must be heated to ignite and have self-sustained combustion independent of the ignition source. If this ignition is caused by an external flame, spark, or object, it is *piloted-ignition*. If the material ignites spontaneously with the increase of temperature, it is *autoignition* or autogenous ignition. For instance, the autoignition temperature of a flammable liquid is the temperature to which a closed or nearly closed container must be heated so that this liquid, when placed in the container, will ignite spontaneously and burn. Normally, the piloted-ignition temperature of a material is considerably lower than the autoignition temperature (1,4).

Knowing the ignition temperature of materials can be extremely important. An illustration of this importance can be seen in the possible exposure of a material, such as carbon disulfide, to a steam line or other heated surface having a temperature of 257°F (125°C) or above, which is the autoignition temperature of carbon disulfide.

As we have seen, fires generally occur where the mixing of fuel and oxygen is controlled by the combustion process itself. On the other hand, explosions generally occur only in situations where the fuel and oxygen have been well mixed prior to ignition. This results in a very rapid combustion reaction, as compared to that of the normal fire. A *deflagration* is an explosion that has gases propagating through the burning material by conduction, convection, and radiation at a rate less than the velocity of sound. A *detonation* is an explosion where a shock wave in the material establishes and maintains the reaction at a rate greater than the velocity of sound.

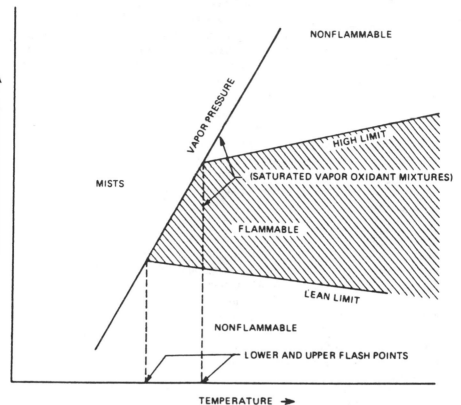

Figure 2.4 Effects of flash point, vapor pressure, and temperature on the limits of flammability of a combustible vapor in air. (Reprinted with permission from the *Fire Protection Handbook,* 15th Edition, Copyright 1981, National Fire Protection Association.)

For a dust explosion, which is basically a deflagration, to occur, the following conditions must be met:

Dust must be present.
An ignition source must be present.
Oxygen must be present in a concentration to sustain rapid combustion.
The dust must be well mixed with the oxygen at a concentration above the lower explosive limit.
Ignition must occur in an enclosed space.

The above fire conditions are also referred to as the *explosion pentagon*.

For an explosion to occur, dust must be well mixed with air both chemically, and physically. For heterogeneous combustion, the rate of reaction is dependent upon the surface area of the dust particles. Small particles may be easily dispersed, ensuring that the maximum available surface area is in contact with the surrounding air. If combustion initiates in this mixture, confinement causes an increase in pressure. The high-pressure gases resulting from the combustion process will try to flow toward a low-pressure area, thereby creating a flow velocity that ensures the mixing of more dust with the air. The rate of combustion increases with increasing pressure, thereby creating even more high-pressure gases and resulting in an explosion. If the requirements for mixing or confinement are not met, a fire rather than an explosion may result (5, 6).

It is possible to prevent an explosion by removing any one of the five conditions. In any loss control program, these conditions must be eliminated or controlled so that a fire and/or explosion will not occur.

Remember that the properties of flammable liquids, gases, and solids are based upon normal ambient temperatures and pressures. A considerable change in these parameters will also have a bearing on the flammable characteristics of the substance located in these different environments. A material under pressure and at high ambient temperatures or in an oxygen-enriched atmosphere can be expected to evidence greater flammability characteristics.

Fire is not a simple chemical-physical reaction. There are many variables, each of which contributes to the behavior of fire. Control methods and the management of fire losses are best accomplished when the role of each variable is understood.

REFERENCES

1. Cote, Arthur E., ed. *Fire Protection Handbook*, 16th Edition, National Fire Protection Association, Quincy, Mass., 1986.
2. *National Fire Codes*, National Fire Protection Association, Quincy, Mass., 1989.
3. *ASTM Standards*, American Society for Testing and Materials, Philadelphia, Pa., 1989.
4. *Loss Prevention Data Books*, Factory Mutual Engineering Corporation, Norwood, Mass., 1989.
5. *Occupational Safety in Grain Elevators and Feed Mills*, U.S. Department of Health and Human Services, Morgantown, W. Va., 1983.
6. *Prevention of Grain Elevator and Mill Explosions*, National Academy of Sciences, National Academy Press, Washington, D.C., 1982.

3
Elements of a Fire Loss Control Program

As new technologies are developed, hazards may be introduced, and when industry and business expand, existing hazards may increase. To remain prosperous, a corporation must minimize the possibility of causalities, interrupted production, and property loss caused by fires and explosions. Corporations are also required by law (OSHA 29 CFR 1910) to provide a reasonably safe workplace for their employees. To satisfy these responsibilities and also remain competitive, corporations must eliminate fire hazards where possible and minimize those that cannot be eliminated (1).

POLICY STATEMENT

A program of identification, evaluation, and control fire hazards is known as a *fire loss control program*. For this program or any program involving the welfare of a corporation or its employees to be effective, the sincere and total support and the sustained interest of top management are essential. This is particularly true for all corporate loss control programs. In addition to top management support, this policy of controlling losses must be communicated to lower levels of management and to all employees. This can be accomplished during the formation of the loss control program by a written policy statement from the chief executive officer demonstrating his or her complete support of the program; outlining the procedures, objectives, responsibilities, and accountabilities; and further indicating his or her desire that all employees of the organization support those responsible

17

for the formation and implementation of the loss control organization and function. This shows that management cares, that management is directly involved.

The objectives of the fire loss control program should be fully stated, with emphasis on the protection of employees against injury and the conservation of corporate assets. Both management and labor benefit from the responsible safeguarding of profit centers and the dependable continuity of operations. This policy should emphasize to all employees that major damage to a facility by fire or explosion in most cases causes temporary or total loss of jobs.

LOSS CONTROL ORGANIZATION

Some form of a loss control organization should be developed to implement any corporate loss control program. No definitive guidelines can be established for the simple reason that such an organization could vary from a very informal structure to one that is highly structured. An informal organization may not involve any full-time personnel, as would be anticipated in relatively small facilities. In this case, a manager may wear two hats—one as the facility/plant manager and one as the loss control manager. As facilities grow in size, corporations find they need full-time personnel in the loss control organization to handle the responsibilities. A large multinational corporation operating numerous divisions and facilities could have plant or division personnel assigned as full-time members of the loss control organization, with a corporate organization providing a staff function.

The old saying, "It takes money to make money," applies indirectly to the development of a loss control organization. Although a dollar return for loss control expenditures may at times be difficult to demonstrate, particularly to a financial staff, it can be unequivocally stated that a loss control organization and program are financially justified. Money must be spent not only to make money, but also to save it. Failure to practice property conservation can result in direct, tangible loss that must be paid for, wholly or in part, with dollars that would otherwise go toward profit. Accidents and other losses can be an intolerable drain on profits.

This does not mean that a large organization cannot have a reasonably effective loss control program without a staff involved full-time in these activities. There are multinational organizations that do not employ fire protection personnel, but assign these tasks to competent and dedicated employees on division or plant levels, or even in individual work areas. Personnel of this type, working with representatives of insurance brokers (who in actuality become the loss prevention organization on a corporate level), can be very effective.

Under any loss control program it is important, as discussed later, to have some control of these functions at the corporate level and to assign line responsibilities at the division and plant levels. Figures 3.1 and 3.2 indicate suggested loss control organizational functions for (1) a large, multidivision, multiplant facility

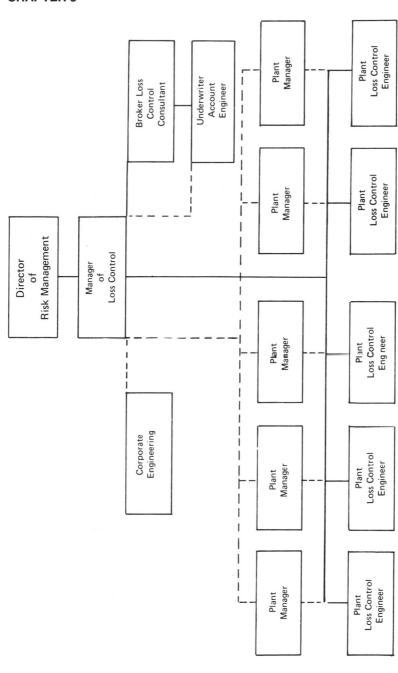

Figure 3.1 A suggested loss control organization for a large multidivision corporation with a corporate loss prevention staff.

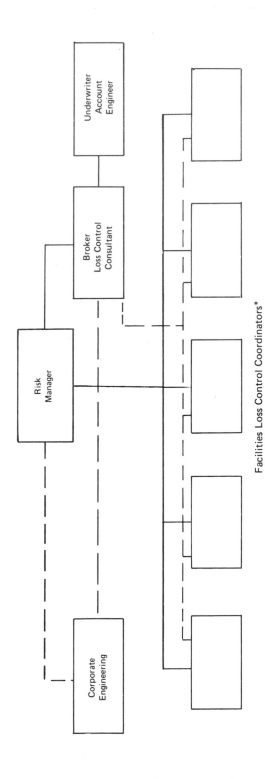

Facilities Loss Control Coordinators*

Figure 3.2 A suggested loss control organization for a large corporation depending principally upon others for loss control efforts. The coordinators (*) will normally have principal responsibilities other than loss prevention. Ideally, coordinators will be members of facilities managers' staffs.

with a corporate loss prevention staff; and (2) functions for a similar organization not employing fire protection personnel, but depending upon outside sources, principally the technical services of insurance brokerage firms.

The loss control manager could be known by a variety of other names depending on the industry—for example, fire loss control manager, fire loss prevention and control manager, or property conservation director. In some industries, the loss control manager reports to the facility manger and is responsible for employee safety management, fire protection management, property insurance management, and security management (1).

Basically, as a minimum, there should be some in-house loss control organization coupled with the assistance of outside sources, such as consulting fire protection engineers, insurance brokerage personnel, and insurance company inspectors and engineers.

IDENTIFICATION

Once a loss prevention organization has been developed, or even if such an organization is not yet in effect, an audit of the facilities should be conducted. This audit is suggested as a first step in identifying hazards so that they can be evaluated and so that priorities can be established for the concentration of efforts in the loss control program. Quite often, the corporate staff will require assistance in the evaluation of their facilities from outside sources. This assistance can be from the insurance personnel, either in a detailed review of property loss prevention reports received from insurance sources with some special visits to particular sites or from technical services personnel of insurance brokers. Brokers will supply this service using either their reports or the services of their control personnel accompanying corporate personnel in the physical inspection of various sites.

An audit form should be developed to be used in the loss control program. A suggested form is shown in Figure 3.3. In addition to identifying the problems at the facility, the audit form can be of considerable value when kept on file at a corporate level for answering questions that frequently develop concerning construction, utilities, various occupancy details, etc.

EVALUATION

Once the information has been collected through the use of the audit program identifying potential fire hazard areas, it is important to develop an improvement

XYZ Corporation
Facilities Audit Form*

Location: Date:

Construction

    Walls: Roof:

    Floors: Partitions:

    Unusual features (combustible interior finishing, insulation, etc.):

Boilers

    Description and rating:

    Fuel:

    Combustion controls (including details on interlocks, purge cycle, flame
       safeguard, etc.):

    Controls testing program:

Electrical

    Power supply (including capacity, no. of feeders, etc.):

    Transformers: No.   Capacity

    Major motors: HP   Spares?

    Hazardous electrical equipment (Class I or II):

    Emergency generators:

    Data processing: Description

        Functions

        Location and envelope construction

        Tape storage

        Air conditioning

Figure 3.3   A form suggested for use in an audit of corporate facilities.

    Protection

    Detectors

  Protection:

    Computer area

    Underfloor

    Tape library

Hazardous operations (flammable liquids, dust, etc.):

Plant protection

  Water supplies (include water test data):

  Underground mains and valving:

  Automatic sprinklers: Full   Partial   Type

    Design

    Alarms

  Special protective systems (CO_2, halon, etc.):

Security

  Watchman service: Alarms:

Environs

  Flood: Seismic zone:

Human-element programs (give date of programs)

  Fire brigade:

  Self-inspections:

  Emergency planning:

  Cutting and welding (hot work):

*Attach plan of facility.

Figure 3.3 Continued

recommendations list based upon the following priorities in descending order of importance:

1.   Life safety exposures: The potential for fire casualties in industrial types of occupancies is directly related to the hazard of the operations or processes. Most of the multiple fatalities are a result of (a) flash fires in highly combustible materials; or (b) explosions in dusts, flammable liquids and vapors, and gases (1, 2, 3). This element of the fire loss control program is covered in greater detail in Chapter 4, Life Safety Elements.

2.   Continuity of operations: This includes hazards in areas that will have a significant effect upon the continuity of operations in the event of their loss, thus affecting possible corporate income and loss of corporate resources. Management is extremely conscious of financial return on investment, and those locations that affect production, such as computer facilities controlling production processes, should be highlighted.

3.   Other hazard areas: This includes problems of protection from hazards not necessarily of a business continuity nature.

Frequently, large firms will need numerous and similar facilities such as distribution locations, construction of theaters, construction of warehouses, etc. Where these situations are encountered, it may be advisable to develop corporate loss control guidelines outlining structural, building services, and protection requirements as related to property loss. The desirability of incorporating guidelines of this nature may be highlighted by an examination of past loss experience, which may show a need for improvement.

While the prime responsibility for the development of the suggested guidelines will rest with the loss control personnel, the project should be coordinated with both operating and facilities personnel. Input and approval should be obtained from all those involved, particularly those having architectural or engineering responsibilities, so that the guidelines will carry the necessary weight and will command the respect and attention of both corporate employees and outside firms who may have some interface with the guidelines.

The components of protection guidelines should include:

1.   Construction details: The construction details include those having a bearing on the structure's ability to withstand a fire or to provide fuel in a fire. This may involve the specific design of a desired fire resistance rating that may be important in guidelines for locations; For examples, specifications for insulated metal-deck roofs so that the roof might be classified as noncombustible rather than combustible; specifying mechanical fastening of the rigid insulation of a metal-deck roof at the roof perimeter to avoid windstorm losses; etc.

2.   Safe installation of building services: This will include the location and

enclosure of switch gear; the location, types, and protection of transformers; the designation of wiring means; the designation of combustion controls for heating and processing equipment and the enclosure of this equipment; the proper installation of flues; the proper installation of air-conditioning systems for sensitive areas such as those housing computers; etc.

3. Required fire protection features: This could take the form of specifying the hydraulic design requirements for a hydraulically calculated sprinkler system for high-piled storage; protection guidelines for computer areas; detection types and engineering requirements for similar installations; for theaters, the protection of storage areas and projection rooms; etc.

When formulating loss control recommendations and guidelines, an important question to ask is: "What loss expectancy can be tolerated?" This is particularly significant when the trend is toward higher insurance deductibles and self-retention of losses. When considering loss expectancy, particularly for lesser valued but numerous locations such as supermarkets or theaters, the desirability of the formulation of loss control guidelines takes on significant importance.

CONTROL

Human Elements

Despite excellent engineering and a good loss control program at the corporate level, problems at facilities and poor loss experience can still exist if the facility manager or first-line supervisor is not involved in the overall effort. This is recognized by insurance underwriters, who insist upon the implementation of what is generally termed human-element recommendations as a basic requirement of coverage. The human elements of a loss control program refer to the continuing actions of personnel at the local level in the loss prevention and control effort in the formulation and updating of emergency planning procedures; in the constant readiness of the emergency organization (fire brigade); in the conduct of self-inspection programs; in the exercise of an established impairment notification system for fire protection system shutdowns; in the use of a cutting, welding, and hot-work permit system; in the limiting of smoking to designated areas; and in good housekeeping procedures. Indeed, these insurance underwriters will frequently live with major protection problems when good human-element policies are established, but it is highly questionable whether they would risk their dollars on long-term clients who will not initiate effective policies of this nature.

A human-element policy is advocated not merely because underwriters attach such importance to its implementation, but primarily because it makes excellent sense. The implementation of human-element programs does not involve extensive expenditures. Such programs provide for greater fire safety and have the further benefit of making local employees more aware of their environment, coupled with a feeling of being a part of the action.

In the following paragraphs, each of the human-element topics will be discussed. However, it is important to remember that it is really not practical to provide an overall human-element program that will fit all facilities.

Emergency Planning

While the need for emergency planning may vary somewhat with different facilities, there should be general agreement on the importance of providing at least some level of planning for emergencies that may be encountered at a particular facility. Previously, the value of an audit of the hazards or problems that could possibly be encountered by local facility managers was discussed. This is the first step in emergency planning: the recognition of the nature of possible emergencies. The second step is determining the impact of these emergencies. Finally, it is important to evaluate the effect that these possible emergencies may have on the interruption or continuity of operations.

Often, reviews of large loss situations indicate confusion and lack of effective actions by local facility personnel at the time of a fire, explosion, or other emergency, which may have added considerably to the adverse impact of disasters. The types of disasters that may need to be addressed should be recognized first. These catastrophes may include, but will not necessarily be limited to, the following:

Fire	Earthquake
Explosion	Civil disturbances
Water damage	Bomb threats
Storms/high winds	Hazardous chemical releases

Once the types of emergencies have been recognized, emergency planning programs should be established to provide a way to address these problem areas properly by emergency response and thus eliminate or lessen any potential loss. The initial actions will involve the emergency organization and outside response. The second action will be the plans that are put into effect after the emergency to lessen its impact by providing ways to continue to provide safety for the employees and occupants and to continue production, maintain the integrity of the facility, etc., by instituting contingency plans.

The third part of emergency planning is to ensure that there is an emergency organization available to implement the plans that have been formulated. This is a very important element of the overall fire loss control program and is covered in Chapter 5, Emergency Organization. The formation of emergency plans must be specific to the individual facility.

While the emphasis for the formation of these plans should be at a local level, corporate management should establish policies requiring the formation of plans and guidelines containing a general outline and areas that should be included in the planning. It may be possible for corporate staff members to provide a general emergency plan if there is a great repetitiveness in locations, which could conceivably be encountered in distribution facilities, a theater chain, or similar types of facilities. The need for emergency planning will, to a great extent, depend upon whether large loss factors exist and upon an analysis of what type of response might be anticipated at the time of a specific emergency.

The emergency plan should be coordinated with any emergency response procedures and hazardous chemical facility profiles that may be required by community emergency planning commissions under the Environmental Protection Agency (EPA) Community Right to Know regulations (40CFR 370). A well-thought-out emergency plan will help plant management to meet the emergency response procedures necessary under the regulations and to ensure that the plant personnel will be protected during a community-wide hazardous chemical disaster.

Elements of emergency planning will consist basically of (1) the audit of catastrophes that could face the particular facilities, and (2) the establishment of the plan itself. Corporate guidelines should outline the areas to be considered in the formation of the plans and should comment on the perils of fire, windstorm, flood, earthquake, and loss of vital services—i.e., power, steam, water, etc. In the first stage, an inquisitive approach is necessary. Asking searching questions will often lead to the discovery of potential areas of extensive loss. On more than one occasion, analysis of losses indicated single failure points that resulted in major escalations of losses. These possible areas should be explored by a very thorough, searching approach.

Generally, middle management will be responsible for recognizing the major loss areas; however, considerable assistance can be obtained from the fire protection personnel of insurance companies, brokers, and consultants. As a result of reviewing many losses, these people have had broad experience and are in a position to identify those areas that may have been missed.

After the formation of the emergency plan, the preparer should critically examine the plan again from a questioning standpoint. The criteria suggested are as follows:

1. Does the plan make provisions for all the possible catastrophes?
2. Are all problems that could conceivably be encountered in each type of disaster anticipated?
3. Does the plan provide for an effective emergency organization to staff the facility during emergencies? This will include the availability of the emergency organization to function on all shifts and off periods.

Self-Inspection Program

Another important part of a fire loss control program is the facility self-inspection. This is recognized by insurance underwriters as being extremely valuable and can help satisfy the OSHA fire protection requirements. The purpose of the self-inspection is basically to ensure that the facility fire protection systems are operable and therefore capable of performing their function; and to detect hazard areas that may create fires, such as poor housekeeping, the unsafe handling of flammable liquids, poorly maintained electrical equipment, and others.

The first item to consider is the assignment of the inspection task. The loss control manager should establish inspection schedules, the types if inspection to be conducted, and the routing of inspection reports. It is also the manager's responsibility to see that these inspections are conducted properly. In a large facility, it is important that the individual delegated with this responsibility have a good mechanical knowledge and, specifically, a knowledge of the facility fire protection equipment. The most logical individual will be maintenance oriented and, in addition, may also be involved in the emergency organization operations. This individual should be reliable, possess the necessary knowledge, and have the confidence of the loss control manager and upper management.

Next, the inspection frequency should be determined. Normally, where sprinkler protection is involved, the recommended frequency is weekly, with particular emphasis on ensuring that water is available and sprinkler control valves are open. Where the facility is of small or moderate size and there are no fixed protection systems, monthly inspections are usually sufficient, since unusual conditions should be readily apparent to facility personnel.

Facility management should make sufficient time available to the inspector to complete the task. The inspector must have confidence that management will correct the deficiencies that he or she finds and give them proper attention on a priority basis.

Fire insurance underwriters usually furnish forms that are generally all-inclusive and that can be used in the performance of self-inspections. On the other hand, the loss control managers at many facilities have found it advisable to adapt a form that is specific to their facility.

As an example, a self-inspection form from one of the major insurance underwriters is shown in Figure 3.4. This type of form is used extensively throughout industry. An example of an inspection form that is specific to a facility is shown in Figure 3.5.

While this section does not cover all possible elements that may be included in an inspection program, some of the items highlighted here are sprinkler system control valves, extinguishers, fire pumps, electrical deficiencies, and the handling of flammable liquids.

Every control valve should be recorded, showing whether each valve is open or closed and sealed or unsealed and including notes about conditions to be remedied. When the seal on a post indicator valve (PIV) is found to be broken or missing, it is advisable to make a drain test at the nearest sprinkler riser to ensure that the valve is open. Even though a target or supervisory switch indicates that the valve is open, PIVs should be tested by turning the valves wide open or until the "spring" is felt, thus ensuring that the valve is in an open position. The outside screw and yoke (OS&Y) valve does not require more than a visual inspection that the valve is open; however, if a seal is broken, the reason for the broken seal should be investigated. With dry pipe valves, checks should be made to ensure that the proper air pressure is provided in the system and, where "quick opening" devices (accelerators and exhausters) are provided, that this equipment is operable. An examination of the sprinkler piping and heads will indicate whether any heads are damaged, corroded, or loaded with deposits, dust, or lint and whether or not sprinklers are obstructed.

Self-inspection programs should include drain tests of sprinkler systems when weather conditions permit. The inspector should know when the normal pressure and record it on the inspection form. This will indicate a problem such as a partially shut valve or other obstructions when the pressure is considerably lower than normal.

Where water supplies depend upon fire pumps, the pumps should be visually checked to determine that power is supplied to the pump where electrical drives are provided, that steam is available where steam is the driving force, and that fuel is available and that batteries are in operation with internal combustion engine drives.

Other parts of the overall facility protection that are also involved in the inspection program include water tanks (properly heated during extremely cold weather), hydrants (not obstructed), hydrant wrenches (available), and hose houses (readily accessible). Accessibility also applies to extinguishes and hose stations within the facility. A check of extinguishers will also indicate whether pressure is satisfactory and whether extinguishers are properly located and serviced.

Part of the inspection program should be to check that proper containers are being used for the handling of flammable liquids, that electrical grounds and bonding are in place, and that bulk quantities of flammable liquids are stored in the flammable-liquids vault.

Unsafe electrical items include overloading and improper overcurrent protection of circuits, poorly maintained electrical cords that are frayed or cracked, combustible storage located within switch-gear rooms, etc.

Other areas involved in the self-inspection program are safe welding and cutting practices, housekeeping practices, fire doors, storage practices, control of

Industrial Risk Insurers

OVERVIEW FORMS PACKET
(See Section 12 in the OVERVIEW Manual)

Overview
A Total Management Program
for Loss Prevention and Control.

FIRE PROTECTION EQUIPMENT INSPECTION REPORT

Facility: _____   Inspector: _____

Location: _____   Date: _____

The Following Items Should Be Checked At Least Weekly.
Any "No" response should be explained.

WATER SUPPLY, SECTIONAL, AND SPRINKLER SYSTEM CONTROL VALVES

Valve ID	Open	Shut	Sealed	Valve ID	Open	Shut	Sealed	Valve ID	Open	Shut	Sealed	Valve ID	Open	Shut	Sealed

PUBLIC WATER

Public water supply in service? ☐ Yes ☐ No _____

Fire department connection accessible, caps in place, couplings free to rotate? ☐ Yes ☐ No _____

FIRE PUMPS

Pump ID	Type	Set For Auto.?		Operated Today?		Checklist Completed?		Pressure: _____ psi Comments
		Yes	No	Yes	No	Yes	No	

WATER SUPPLY TANKS

Tank ID	Tank Full?		Heater Working?		Water Temp.	Comments
	Yes	No	Yes	No		

AUTOMOTIVE FIRE APPARATUS

Each fully in service? ☐ Yes ☐ No _____

Checklist completed? ☐ Yes ☐ No _____

SPECIAL EXTINGUISHING SYSTEMS

System ID	Type	In Service?		Date Last Serviced	Date Last Tested	Comments
		Yes	No			

The Following Items Should Be Inspected At Least Monthly.

Any "No" response should be explained.

WET PIPE, DRY PIPE, DELUGE, AND PRE-ACTION SPRINKLER SYSTEMS

System ID	Alarm Tested?		Water Pressure			Heat Adequate?		Air/ Supv. Press.	Comments
	Yes	No	Static	Flow	Differ-ential	Yes	No		

Figure 3.4 Example of a self-inspection form. (Reproduced with permission of Industrial Risk Insurers.)

FIRE EXTINGUISHERS, INSIDE HOSE CONNECTIONS, AND STANDPIPES

Each unit in service? ☐ Yes ☐ No _____

Checklist completed? ☐ Yes ☐ No _____

HYDRANTS, HOSE HOUSES, AND MONITOR NOZZLES

Monitor Nozzle/ Hydrant ID	Accessible?		Drained?		Equipment				Comments
					Adequate?		Cond. OK?		
	Yes	No	Yes	No	Yes	No	Yes	No	

FIRE DOORS

Fire doors and shutters in good condition? ☐ Yes ☐ No _____

Automatic closing devices operable? ☐ Yes ☐ No _____

SMOKE AND HEAT, AND EXPLOSION-RELIEF VENTS

Vents operable? ☐ Yes ☐ No _____

Areas around vents unobstructed? ☐ Yes ☐ No _____

PROTECTIVE SIGNALING SYSTEMS

All systems been tested satisfactorily? ☐ Yes ☐ No _____

OTHER PROTECTION DEFICIENCIES FOUND DURING THE COURSE OF EACH INSPECTION SHOULD BE REPORTED BELOW:

If "Yes," note location.

	Yes	No
Stock within 36 inches of sprinkler heads?	☐	☐
Sprinkler heads or piping bent?	☐	☐
Sprinkler heads painted?	☐	☐
Sprinkler heads or piping corroded?	☐	☐
Sprinkler heads loaded with debris?	☐	☐
Items hanging from, or supported by sprinkler heads?	☐	☐
Sprinkler heads obstructed by partitions?	☐	☐
Signs of internal sprinkler piping obstruction?	☐	☐
Fire doors blocked by materials?	☐	☐

ADDITIONAL COMMENTS AND RECOMMENDATIONS

Report reviewed by: _____   Position: _____
                            (signed)

Has prompt action been initiated?   ☐ Yes   ☐ No

FILE FOR REVIEW BY IRI REPRESENTATIVE

Figure 3.4   (Continued) Back of form.

DISTRIBUTION: ORIGINAL TO: FACILITIES SERVICES
                 COPY TO: INSURANCE DEPARTMENT
If condition is satisfactory, enter a check ☑ . If unsatisfactory enter ☐O☐ and explain on back of this form.

		B	1	2	3	4	5	6	
Housekeeping	General orderliness & cleanliness								
Portable Fire Extinguishers	Water units tagged within 12 mos.								
	CO_2 & dry chem. units tagged within 6 mos.								
	Gage pressure in operating range								
	Properly mounted & sealed								
	Accessible								
Exits	All aisles & doors clear								
	Doors open freely								
	No storage in stairways								
	No stairway doors blocked open								
	All exit signs & stairway lights lit								
Fire Hose	In rack								
	Connected								
	In good condition								
Fire Doors	None blocked open								
	No damaged doors								
	All operate freely								
	Fusible links missing or painted								
Electrical	Junction & panel boxes closed								
	No temporary wiring								
	No frayed or unsafe wiring								

Sprinkler Control Valves	Open	Locked
2½" O.S. & Y.'s in 1st floor rear passageway		

Prepared by: _____   Mgmt. review by: _____   Deficiencies Corrected ☐ ☐
             Name   Title   Name   Title   Yes No
                         *IF "NO", EXPLAIN ON BACK OF SHEET

Figure 3.5 A self-inspection form specific to a facility.

smoking, heating, water supplies, special hazards, and special types of protection. These items, along with the previously mentioned items, will be covered in other sections.

Permit Systems

Hot Work   Cutting and welding cause about 6% of the fires reported in industrial properties (4); consequently, these activities should be carefully controlled and safeguarded. Hot-work permit systems, including cutting and welding or other use of open-flame or spark-producing equipment, are the rule rather than the exception in industrial facilities. This type of system should be adopted throughout industry, but is absolutely critical in certain facilities involving the presence of flammable vapors and liquids or in combustible dust locations.

The permit system is simply a system that uses a tag or form to permit these operations in a specified area. The permission is usually granted by management for a specified period of time and requires that the site be checked prior to the granting of permission. A sample cutting and welding permit tag, supplied by a major insurer, is shown in Figure 3.6. An example of a permit form is shown in Figure 3.7. The objective of the permit system is basically to provide a safe area for these operations and to ascertain that fire protection is available. The hazards and control of hazards of cutting and welding operations are covered in Chapter 7, Common Process Hazards.

Confined Spaces   Most confined spaces are not designed for employees to enter and perform work. The spaces are typically used for product storage, enclosing processes, or transporting products. Entry is often difficult and dangerous due to the location and size of the openings, lack of air movement, and potential hazards. In addition to hazardous atmospheres such as toxic, oxygen-deficient, irritant, and corrosive, flammable atmospheres maybe encountered in a confined space. A flammable atmosphere generally arises from oxygen enrichment, vaporization of flammable liquids, by-products of work, chemical reactions, concentrations of combustible dusts, or desorption of chemicals from inner surfaces of the space.

As with hot work, entry into confined spaces should be by permit only. This permit is a written authorization and approval that specifies the location and type of work to be done and certifies that all existing hazards have been evaluated by a qualified person designated by management and that necessary protective measures have been taken to ensure the safety of employees (5, 6). A sample permit is shown in Figure 3.8.

Notification of Impairment of Fire Protection Equipment   The policy of providing this impairment notification, also known as fire protection system shutdown, and the actions to be taken in the event of impairment, should be clearly stated at the corporate level. In the vent of a major impairment, extra precautions should be taken in the form of a fire watch; additional watchman service; the laying of fire

Cutting/Welding
Permit
Applies Only to Area
Specified Below

SEE INSTRUCTIONS ON REVERSE SIDE

SECTION A	
LOCATION	JOB NO.

SECTION C	
DATE	JOB NO.
LOCATION & BUILDING	FLOOR
NATURE OF JOB	
WELDER'S NAME	

The above location has been examined. The precautions checked
on the reverse of this card have been taken to prevent fire.
Permission is granted for this work.

PERMIT EXPIRES:	DATE	TIME	AM PM

SIGNED (FIRESAFETY SUPERVISOR)

215 (1-84)PUBS   PRINTED IN USA

Figure 3.6   A sample cutting and welding permit tag. (Reproduced with permission of Factory Mutual System.)

WATCH FOR FIRE

Cutting & Welding Recently Done Here

SECTION B		
LOCATION		JOB NO.

AREA CHECKED (2-4 Hours after work completed)	TIME OF PICK-UP	AM PM	DATE
	SIGNATURE OF PERSON RESPONSIBLE		

DATE	JOB NO.
LOCATION & BUILDING	FLOOR

NATURE OF JOB

WELDER'S NAME

The above location has been examined. The precautions checked on the reverse of this card have been taken to prevent fire. Permission is granted for this work.

PERMIT EXPIRES:	DATE	TIME	AM PM
SIGNED (FIRESAFETY SUPERVISOR)			

TIME STARTED	AM PM	TIME FINISHED	AM PM

FINAL CHECKUP BY WELDER

Work area and all adjacent areas to which sparks and heat might have spread (such as floors above and below and on opposite side of walls) were inspected after the work was completed and were found firesafe.

Signed: _____

After signing, return permit to person who issued it.

Figure 3.6 (Continued)

Sample Fire Permit Form

Permit Number _____　　Date _____
Area _____　　　　　　Time _____

Precautions	Yes	No	Not Applicable
Atmosphere tested	☐	☐	☐
Spark-proof tools issued	☐	☐	☐
All combustible material moved	☐	☐	☐
Flame-proof tarps or covers in use	☐	☐	☐
Welding area enclosed	☐	☐	☐
Shield screens in use	☐	☐	☐
Standby fire watch*	☐	☐	☐
Ventilation checked	☐	☐	☐
Purge line with inert gas	☐	☐	☐
All openings closed to prevent spread of sparks	☐	☐	☐

Fire Equipment on Site
Type of Equipment _____
　　　　　　　　　_____
　　　　　　　　　_____

Additional Precautions:
_____
_____
_____

*Always required for welding outside maintenance shop.

_____
Maintenance Supervisor's Signature

_____
Production Supervisor's Signature

Figure 3.7　A sample fire permit form. (Courtesy of North Carolina Department of Labor.)

hose; and, most importantly, the alerting of all personnel that extreme precautions should be taken to avoid the possibility of fire during the impairment period.

The most desirable impairment notification system will involve the use of a three-part tag, with one part of the tag attached to the system indicating that the system is impaired. This tag will be *red*, so that it is readily identifiable. One portion of the tag is forwarded to the insurance underwriter or to others. The third portion is retained to indicate the correction of the impairment. A sample of this type of notification tag system used by an insurer is shown in Figure 3.9. Other systems, however, may be equally effective and may involve the immediate notification of personnel involved in loss prevention by phone, followed by written notice, with a follow-up indicating the rectification of the impairment problem. This type of impairment notification form is shown in Figure 3.10. The importance of follow-through on all impairments cannot be overemphasized if a property conservation or loss prevention program is to be effective.

Good Housekeeping Practices

Principles of Good Housekeeping Housekeeping practices, or the maintenance of orderly cleanliness and neatness, are basic to a good fire loss control program. These practices reduce the danger of fire loss because they are proven methods of controlling the presence of unwanted fuels, obstructions, and sources of ignition. Housekeeping practices consist of the simpler aspects of building care and maintenance, proper control of waste, and regulation of personnel practices (i.e., control of smoking). The basic requirements of good housekeeping can be generalized in these areas:

Proper layout and equipment
Correct materials handling and storage
Cleanliness and orderliness

Sufficient working space, adequate storage areas, and proper equipment for moving materials are necessary to keep work areas from becoming obstructed. Improper storage of materials can block exists and access to important fire protection equipment. Keeping all areas clean reduces the threat of dangerous accumulations of materials and trash (1,7).

Control of Smoking Uncontrolled smoking still ranks high on the list of causes of industrial fires (18% of industrial fire origins are due to smoking of tobacco) (8). While the need to control smoking is obvious, it should be stated in the overall corporate fire loss control policy that smoking *will* be controlled within the plant premises and confined to safe, designated locations. Industrial Risk Insurers, in its publication *Ten-Point Pocket Guide To Fire Protection*, provide 10 rules for controlled smoking (9). These are as follows:

1. Provide safe, supervised, convenient areas where smoking is permitted.
2. Equip sprinklered smoking room with ample "butt" receptacles and ample extinguishers; keep clean and free of combustibles.
3. Maintain a continuous campaign against reckless smoking and careless use of matches.
4. Designate restricted areas and mark them clearly with "No Smoking" signs.
5. Use only safety matches or electrical or mechanical lighters.
6. Crush all "butts," killing the glow, and discard in a safe place.
7. Enforce the "No Smoking" rule in restricted areas.

SAMPLE PERMIT

CONFINED SPACE ENTRY

CLASS ____

Location of Work: _____

Description of Work (Trades): _____

Employees Assigned: _____

Entry Date: _____ Entry Time: _____

Outside Contractors: _____

Isolation Checklist:

Blanking and/or Disconnecting
Electrical
Mechanical
Other

Hazardous Work:

Burning
Welding
Brazing
Open Flame
Other

Hazards Expected:

Corrosive Materials
Hot Equipment
Flammable Materials
Toxic Materials
Drains Open
Cleaning (Ex: chemical or water lance)
Spark Producing Operations
Spilled Liquids
Pressure Systems
Other

Vessel Cleaned:

Deposits _____
Method _____
Inspection _____
Neutralized with _____

Fire Safety Precautions: _____

Figure 3.8 A sample permit form for confined space entry. (Courtesy of National Institute of Occupational Safety and Health.)

Personal Safety:

    Ventilation Requirements
    Respirators
    Clothing
    Head, Hand, and Foot Protection
    Shields
    Life Lines and Harness
    Lighting
    Communications
    Employee Qualified
    Buddy System
    Standby Person
    Emergency Egress Procedures
    Training Sign Off (Supervisor or Qualified Person)_____
    Remarks: _____

Atmospheric Gas Tests

      Tests Performed   -   Location   -   Reading

Example: (Oxygen)_____   _____   (19.5%)_____
Example: (Flammability)   _____   (Less than 10% LFL)
         _____   _____   _____
         _____   _____   _____
         _____   _____   _____

Remarks: _____

Test Performed By: _____
                            Signature

Time: _____

Authorizations:

    Supervisor: _____
    Prod Supervisor: _____
    Line Supervisor: _____
    Safety Supervisor: _____
    Etc.: _____

Entry and Emergency Procedures Understood:

    Standby Person _____
    Rescue _____
    Telephone _____

Permit Expires: _____

Classification: _____

Industrial
Risk
Insurers # FIRE PROTECTION SHUT OFF
ATTACH TO VALVE OR DISCONNECTING DEVICE

SHUT OFF: _____ , _____
                              DATE   TIME

BY: _____

AUTHORIZED BY: _____

FOR VALVE CLOSURE:

TURNS TO CLOSE: _____ TURNS TO OPEN: _____

DRAIN TEST:

_____ PSI STATIC — _____ PSI FLOWING = _____ PSI

After valve has been opened, match this tag with the office reminder and file until next
Industrial Risks Insurers' Inspection.

NOTIFY IRI OF IMPAIRMENT

(Form N-232-Rev. 4/87)   Printed in USA

Industrial
Risk
Insurers # IMPAIRMENT REMINDER
THIS CARD SHOULD BE DISPLAYED IN A VISIBLE LOCATION
UNTIL SHUT OFF TAG IS RETURNED

PROTECTION: ☐ SPRINKLERS ☐ FIRE PUMP ☐ TANK ☐ CITY WATER
☐ UNDERGROUND ☐ ALARM SYSTEM ☐ SPECIAL EXT. SYSTEM.

VALVE NO. _____ LOCATION _____

REASON: _____

SHUT OFF: _____ , _____
                              DATE   TIME

BY: _____

AUTHORIZED BY: _____

NOTIFY IRI OF RESTORATION

RESTORED: _____ , _____
                           DATE   TIME

Figure 3.9   A sample impairment notification tag system. (Reproduced with permission of Industrial Risk Insurers.)

FIRE PROTECTION SYSTEM
SHUTDOWN REQUEST

Emergency_____   Routine_____

Type of Utility or Utilities to be Shutdown_____

_____

Reason for Shutdown_____

_____

Fire Watch Req'd_____   Number_____

Location of Shutdown_____

_____

Type of Shutdown:

Valve___ Zone___ SD___ HD___ Bag___ Shunt___   Total_____

Length of time down_____   What day(s)?_____

Hours: Start at_____   Back on_____   Date on_____

Shutdown requested by: Contractor_____   FM_____

Person requesting shutdown:_____

Authorized signature, FM Area Manager_____

Authorized EH&S signature_____ or OK by phone___

Note:_____   Job Number_____Acct. No._____

Comments:_____

_____

Cleared by:_____   Time:_____   Date: / /
Note: All shutdowns require at least a 24-hour notice, except in case of emergency.

Figure 3.10   Example of an impairment notification form.

8. Develop safe personal smoking habits.
9. See that visitors are informed of the "No Smoking" rule.
10. Make sure all outside contractors observe these regulations.

Since smoking is also considered a major health threat in addition to being a possible ignition source, a number of communities have passed ordinances limiting where smoking can take place in both public and private buildings. This should be factored in when considering a controlled smoking program. Obviously, the intent of the control of smoking is to recognize that workers will smoke and to provide them with ways to satisfy this need in safe locations that are designed for this purpose, have proper receptacles, and are free of combustibles.

Fire Prevention Plan

With any discussion of a fire loss control program, it is important to ensure that any local, state, or federal regulations in the areas if fire prevention and protection are followed. OSHA has established a number of general industry regulations, including fire protection requirements covering means of egress, fire brigades, fixed and portable fire suppression equipment, fire detection systems, fire prevention plans, and emergency action plans (10). To help in the management of fire loss control, the intent of these requirements will be covered under the appropriate chapters.

This section discusses a model plan to prepare for emergencies and fire prevention and outlines the specific OSHA criteria to follow (including references to OSHA regulations). This plan was adapted from the model plan outlined in the North Carolina Department of Labor's *A Guide to Occupational Fire Prevention and Protection* (8). By auditing the work area, by training the employees, by acquiring and maintaining the necessary equipment, and by assigning responsibilities and preparing for an emergency, human life and facility resources will be preserved. At the beginning, this model plan requires management decisions as to whether employees will be employed to fight fires. Management's selection of a course of action regarding employees and fire protection depends on the requirements and the needs of each individual facility. This decision, usually made by top management, requires careful consideration. The most important factors in providing adequate safety in a fire are the availability of proper exit facilities to ensure ready access to safe ares and the proper education of employees as to the actions to be taken in a fire (8).

There are two basic options available to management:

Option A: Employees will fight fires.
Option B: Employees will not fight fires.

OPTION A: WILL FIGHT FIRES

The selection of Option A entails two additional decisions. First, who will fight fires: (1) all employees, (2) designated or selected employees, (3) fire brigade/emergency organization, or (4) any combination of these. Second, what type of fires will be fought: (1) incipient stage fires only, or (2) interior structural fires.

Once these decisions have been made, management should follow one of the following plans:

Plan 1 All employees. Provide education in fire extinguisher use and hazards involved with incipient stage fire fighting upon initial employment and annually thereafter (29 CFR 1910.157 {g}) (10). Should an employer expect all employees to fight an incipient stage fire in their immediate work areas, as a designated group (such as a fire squad), those employees should receive hands-on training with the appropriate fire-fighting equipment.

Plan 2 Designated or selected employees. Provide education in fire extinguisher use and hazards involved with incipient stage fire fighting upon initial employment and annually thereafter. Provide an emergency action plan that designates specific employees to use fire-fighting equipment (29 CFR 1910.38{a}) (10). Provide annual hands-on training with the appropriate fire-fighting equipment.

Plan 3 Fire brigade (emergency organization) (incipient stage fires only). Prepare a fire brigade organizational statement that establishes the existence of a fire brigade (organizational structure, training, number of members, functions) (29 CFR 1910.156) (10). Provide training for duties designated in organizational statement. Provide hands-on training annually in the use of extinguishers, 1-1/2-in. hose lines (such as Class II standpipe system), and small hose lines (5/8 in. to 1-1/2 in.). Train and educate members in special hazards and provide standard operational procedures. Provide a higher level of training and education for leaders and instructors.

[Or] Fire brigade (emergency organization) (interior structural fires). Prepare a fire brigade organizational statement. Ensure physical capability of members. Provide training for duties assigned in fire brigade organizational statement. Provide educational or training sessions at least quarterly and hands-on training with appropriate fire-fighting equipment at least annually. Train members in special hazards and provide standard operational procedures. Provide higher level of training and education for leaders and instructors. Provide required protective clothing and breathing apparatus.

Option B: Will Not Fight Fires

The selection of Option B entails one additional decision—whether to have portable fire extinguishers and hoses in the facility (unless a specific regulation requires that extinguishers be provided).

One of the following plans should be followed:

Plan 1 Provide fire extinguishers and hoses. Provide fire extinguishers and hoses required by regulation or insurance carrier. provide emergency action plan and fire prevention plan (29 CFR1910.38{a} and {b}) (10). Maintain and test this equipment. Provide for critical operations shutdown and evacuation training.

Plan 2 Fire extinguishers and hoses not provided. Provide emergency action plan and fire prevention plan. Provide for critical operations shutdown and evacuation training.

Elements of Fire Prevention Plan

As was seen earlier, depending on what option is chosen, a written fire prevention plan may be required. Obviously, if a good fire loss control program is established, the required elements of the fire prevention plan will be included automatically. If there is no fire loss control program, then at a minimum a fire prevention plan should be developed. Facility managers and/or loss control managers should be responsible for developing this plan and keeping it current. This section discusses the necessary elements of the plan.

Element 1 **Names of persons responsible for control of fire protection equipment and ignition sources.** Persons who are responsible for the control and maintenance of equipment related to fire control or for the control of particular hazards should be clearly identified. An example form for personnel assignments is shown in Figure 3.11.

Element 2 **Control of major workplace fire hazards.** Major fire hazards peculiar to the facility and a plan for control of such hazards should be identified. The plan should include proper handling and storage procedures, potential ignition sources and their control procedures, and the type of fire protection equipment available. An example of a type form that could be used is shown in Figure 3.12.

Element 3 **Housekeeping procedures established.** Proper housekeeping is an element in ensuring effective fire prevention.

Element 4 **Fire prevention for heat-producing equipment.** Heat-producing equipment, such as heaters, furnaces, and temperature controllers for such equipment, need maintenance to ensure proper operation. An example form for keeping track of heat-producing equipment and controls is shown in Figure 3.13.

Element 5 **Review plan with employees**. Once established, the plan should be reviewed by all employees covered by the plan to ensure that they are aware of the types of fire hazards of the materials and processes to which they might be exposed.

Elements of Emergency Action Plan

Elements of an emergency action plan are covered under the "Emergency Action Plan" section of Chapter 4, Life Safety Elements.

Fire Protection and Prevention Assignments

Name	Work Location	Job Title	Assignment
_____	_____	_____	Responsible for maintenance of equipment and systems installed to prevent or control ignition of fires.
_____	_____	_____	Responsible for control of fuel source hazards.
_____	_____	_____	Responsible for regular and proper maintenance of equipment and systems installed on heat-producing equipment to prevent fires.

Emergency Plan and Fire Protection Plan Coordinator:

Name: _____ Date: _____

Figure 3.11 Example of a form for personnel assignments. (Courtesy of North Carolina Department of Labor.)

Control of Major Workplace Fire Hazards

The following major fire hazards are present in the work area, and proper control procedures are described.

Major Fire Hazard	Location	Company Controls
1. _____	_____	_____
2. _____	_____	_____

Figure 3.12 Example of a form for listing fire hazards and control procedures. (Courtesy of North Carolina Department of Labor.)

Fire Prevention for Heat-Producing Equipment

Heat-Producing Equipment and Controls	Routine Maintenance	Frequency of Maintenance	Assigned Responsibility
_____	_____	_____	_____

Figure 3.13 Example of a form for listing heat-producing equipment and controls and maintenance procedures. (Courtesy of North Carolina Department of Labor.)

REFERENCES

1. Cote, Arthur E., ed., *Fire Protection Handbook*, 16th Edition, National Fire Protection Association, Quincy, Mass., 1986.
2. Coté, Ron, "Life Safety in Industrial Occupancies", *Industrial Fire Hazards Handbook*, 2nd Edition, Jim L. Linville, ed., National Fire Protection Association, Quincy, Mass., 1984.
3. Lathrop, James K., ed., *Life Safety Code Handbook*, 4th Edition, National Fire Protection Association, Quincy, Mass., 1988.
4. *National Fire Codes*, National Fire Protection Association, Quincy, Mass., 1989.
5. *Working in Confined Spaces*, U.S. Department of Health and Human Services, Morgantown, W.Va., 1979.
6. Pettit, Ted and Linn, Herb, *A Workers Guide to Confined Spaces*, National Institute for Occupational Safety and Health, Morgantown, W.Va., 1984.
7. Shaw, Deborah A., ed., *NFPA Inspection Manual*, 6th Edition, National Fire Protection Association, Quincy, Mass., 1989.
8. Smith, Michael R, ed., *A Guide to Occupational Fire Prevention and Protection*, NC-OSHA Industry Guide #4, North Carolina Department of Labor, Raleigh, N.C., 1989.
9. *Ten-Point Pocket Guide to Fire Protection*, Industrial Risk Insurers, Hartford, Conn., 1979.
10. *General Industry Standards—29 CFR1910*, Occupational Safety and Health Administration, Washington, D.C., 1988.

4
Life Safety Elements

The most important element of a fire loss control program is the life safety of building occupants and employees. In today's business and industrial world, the safety and health of the employee are emphasized more than ever before. This emphasis also has been carried to the other occupants of the building and to the community in general. The federal government, through OSHA, has established a number of general industry regulations (1). Two of these, the fire protection requirements and the hazard communication requirements, deal with life safety and fire protection for employees in industrial occupancies, both manufacturing and nonmanufacturing. The fire protection requirements cover means of egress, fire brigades, fixed and portable fire suppression equipment, and fire detection systems, along with fire prevention plans and emergency action plans. The hazard communication requirements, often referred to as "right-to-know" laws, cover not only health, but also physical hazards—fire, reactivity, pressure, explosivity—of materials that employees can be exposed to.

Through the EPA, the federal government has also established a number of regulations concerning the community's right to know of the processing, storing, manufacturing, and use of hazardous materials by industry. These are similar to OSHA's hazard communication plan, but on the level of the community affected by the industries (2). These standards were developed as a direct result of the recent disasters such as Bhopal, India, toxic chemical release in which at least 2000 people were killed and about 200,000 were injured. Over the years, other major releases—such as BLEVEs (boiling liquid expanding vapor explosions) and derailed train cars—have occurred that in some cases were both a fire and an explosion hazard.

Thus, the fire loss control program not only needs to cover the life safety of employees, but also may have to cover certain elements that relate to the surrounding community. Some of these elements, in particular the OSHA fire protection requirements and fire prevention plan, will be covered in other chapters; the emergency action plan requirements will be covered later in this chapter.

LIFE SAFETY CONCEPT

Understanding the life safety concept is important as a first step in looking at the safety element in a fire loss control program. This will entail a brief general discussion of life safety factors, employee or occupant factors, the nature of fire as it related to life safety, and life safety principles.

Life Safety Factors

To ensure life safety is to ensure that people are not exposed to fire, smoke, and other products of combustion. The magnitude of risk is determined by the development of the fire and the susceptibility of people to the fire. As a fire develops over a period of time, heat and smoke build up to create an environment that is a threat to life safety. Therefore, time becomes a critical factor.

After ignition, most fires begin slowly, but at some point the environment starts to deteriorate rapidly. Figure 4.1 gives a general approximation of how a fire may develop to a critical level (3). As shown in the figure, there is some degree of deterioration prior to detection. The critical level is the point at which the environment has deteriorated to a stage that is a threat to life. This occurs at a certain time interval from the ignition point. The time between detection and the critical point is the time available for evacuation, automatic fire suppression, and other actions necessary so that people will not be exposed to the critical level of deterioration. Obviously, if the fire is detected more quickly, at a lower detection level, the time available for action is increased. At the same time, fires with different slopes to their development curves will increase or decrease the time available for action. For example, as the slope of the curve decreases, representing a slower developing fire, the action interval increases. This can be seen in Figure 4.2 where curve B depicts a slower developing fire than does curve A (3).

Occupant Factors

The occupants of the building are the population at risk. A number of factors may affect their ability to escape from a fire: age, mobility, awareness, knowledge, and control of occupants.

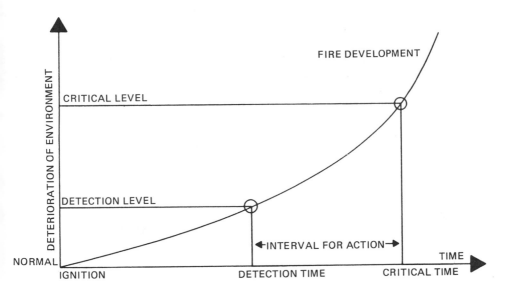

Figure 4.1 Approximate rate of deterioration of the environment. (Reprinted with permission from the *Fire Protection Handbook,* 15th Edition, Copyright 1981, National Fire Protection Association, Quincy, MA 02269.)

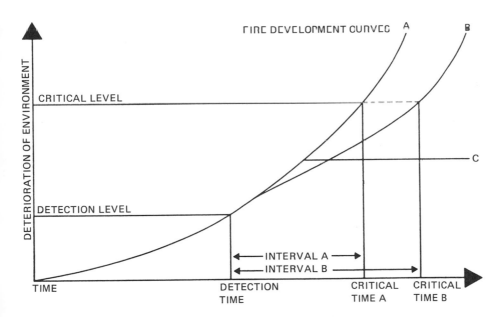

Figure 4.2 Differences in the action intervals between a faster developing and a slower developing fire. (Reprinted with permission from the *Fire Protection Handbook,* 15th Edition, Copyright 1981, National Fire Protection Association, Quincy, MA 02269.)

Age is the easiest to identify and may have a direct impact on mobility, awareness, knowledge, and control. For the most part, the very young and very old suffer the highest fire fatality rates.

Mobility is basically a function of age and physical capability. Building designs today incorporate access for disabled persons. These requirements are written into the local and national regulations. The limitations of disabled persons must be factored into the assessment of the life safety risk. Mobility can also be restricted in occupancy types such as hospitals, psychiatric institutions, or prisons (3, 4).

People can be awake or alert, as in business or industrial settings, or asleep, as in residential or hotel settings. Awareness can also be affected by alcohol or narcotics.

Knowledge can be increased through training and emergency evacuation drills. A person who is familiar with a building and its exiting arrangement has a greater chance of escaping a fire in the building than does a transient visitor (3, 4).

Density and control of occupants are characteristics of the occupants as a group rather than as individuals. The density of occupants is simply the number of people in an area of the building. Density plays an important role in assessing the life safety risk. The higher the density, the more likely the speed of evacuation will be decreased. If there are more people in an area, there also is a greater potential for injury. Control of occupants relates heavily to training and evacuation drills. Industrial occupancies with an emergency evacuation plan and experienced fire wardens will have greater control than occupancies without this type of control, such as a theater or restaurant. With control, there is a decreased possibility of a panic situation, since people can respond in a systematic way to an emergency (3).

Nature of Fire

The factors relating to the nature of the fire that most affect the life safety of people are the ignition potential, the growth of the fire, and the spread of fire and smoke.

Ignition consists of a heat source in contact with or close to a fuel source. In buildings, there are a number of different heat sources (e.g., energy sources, process heat, friction) and fuel sources (e.g., wall coverings, furnishings, flammable liquids and gases). Time is also an important factor in ignition, since the exposure time of the heat source to the fuel source impacts on whether there is ignition or not. A high-temperature source for a short duration might not ignite, where a lower temperature source for a longer duration will ignite. This can be seen in Figure 4.3 (3).

Fire growth is critical. The rate of the fire growth impacts on decisions and actions taken to protect the occupants and to detect and suppress the fire. Al-

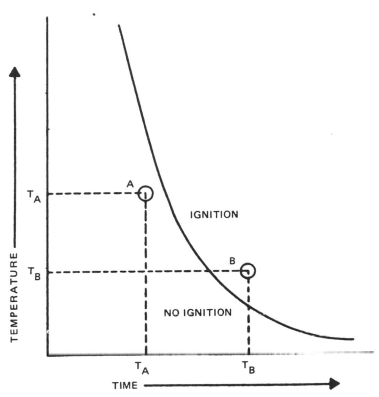

Figure 4.3 Rate of ignition at different temperatures. (Reprinted with permission from the *Fire Protection Handbook,* 15th Edition, Copyright 1981, National Fire Protection Association, Quincy, MA 02269.)

though the immediate fuel is important in the first stages of a fire, the surrounding environment becomes more important in the later stages. Fuel burning near a corner has a faster rate of fire growth than does fuel burning in the middle of the room, because the wall heats up and radiates heat back to the fuel with less heat loss from the original burning fuel (see Fig. 4.4). The ceiling height is another important factor in fire growth. A fire will spread much farther if the ceiling is low, because the flame will "mushroom" out over the ceiling (Fig. 4.5) rather than growing vertically as it will with a high ceiling (5). This type of flame spread causes faster ignition of other fuel or materials in the area. Once this occurs, the surrounding walls and ceiling will begin to rise in temperature until reaching equilibrium with the original fire. At this point, the oxygen level drops sharply and large amounts of carbon monoxide, carbon dioxide, and other gases are liber-

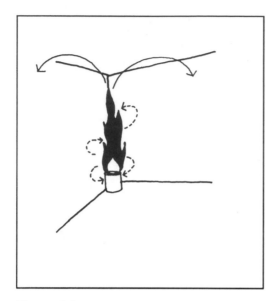

Figure 4.4   Effects of fuel location on fire growth.

Figure 4.5   Effects of ceiling height on fire growth.

ated. In a brief period, normally less than a minute, the temperature can rise several hundred degrees and the flames can spread throughout the space. This phenomenon is known as *flashover*. At this point, human survival is not possible in the room of origin of the fire (5).

The spread of fire and smoke affects people beyond the room where the fire originates. Due to a positive pressure created in the area of fire origin, smoke and heat tend to be distributed to other areas of the building. Smoke and heat travel through the same areas used by people when evacuating a fire—such as open doors, corridors, and open stairwells. The smoke also travels by way of building features, such as plenums, ducts, shafts, and openings around pipes (3).

Life Safety Principles

The *Life Safety Code*, NFPA Std. 101, has been adopted by numerous municipalities, as well as by the federal government in its OSHA regulations (6). It is also used as a guide to good life safety practice. This code, which is concerned primarily with life safety, differs from building codes, which cover both life safety and preservation of property.

Generally, life safety is based on the following principles, which are also covered in the *Life Safety Code* (3, 6):

Provide a sufficient number of unobstructed exits.
Protect exits from growth
Provide alternate exits.
Subdivide areas and construct to provide safe areas of refuge in occupancies
    where evacuation is a last resort.
Protect vertical openings to limit the spread of fire.
Ensure early warning of fire.
Provide adequate lighting.
Ensure that exits and routes of escape are clearly marked.
Provide emergency evacuation procedures.
Ensure that construction is adequate to provide structural integrity during a fire
    while people are evacuating.

The *Life Safety Code* deals with occupancy groups according to their safety hazard. These groups are (1) assembly, (2) educational, (3) health care and penal, (4) residential, (5) mercantile, (6) business, (7) industrial, (8) storage, and (9) unusual structures. Each group has life safety requirements specific to that occupancy.

When discussing the hazards of contents, the three classifications are low, ordinary, or high (3, 6). *Low hazard* covers contents with low combustibility, such that no self-propagating fire can occur, with the only probable danger being from panic, fumes, smoke, or fire from an external source (6). *Ordinary hazard* covers contents that are liable to burn with moderate rapidity or to develop a considerable volume of smoke, but with no poisonous fumes or threat of explosion (6). *High hazard* covers contents that are liable to burn with extreme rapidity with poisonous fumes and the threat of explosion (6).

Some of the life safety features of industrial facilities will be discussed in the following paragraphs. A more detailed discussion of the life safety requirements of industrial facilities and other occupancies can be found in the *Life Safety Code* (6).

LIFE SAFETY IN INDUSTRIAL FACILITIES

Every year the industrial property destroyed by fire makes up a major percentage of the total property loss. Fortunately, there is not a similar percentage of total life loss from fires (3). Around 4% of the fire fatalities occur in the occupational setting. More importantly, employee fire and explosion casualties, as a single source of injury category, represent only about 5% of the total occupational fatalities (7). The low fire casualties are generally due to the operating and protection features and characteristics of the occupants typically found in industrial facilities.

One of the most important factors in this success has been the automatic sprinkler system. Originally developed for property protection, the automatic sprinkler can control the spread of fire, giving employees more time to evacuate. The sprinkler water flow alarm also can alert employees to the fire, which improves reaction time (3, 4, 8). Automatic sprinkler systems are now being designed for life safety with the advent of "quick-opening heads" and other features that make the automatic sprinkler more conducive to life safety protection.

The potential for fire casualties in this type of occupancy is directly related to the hazard of the operations or processes. Most of the multiple fatalities are a result of (1) flash fires in highly combustible materials, or (2) explosions in dusts, flammable liquids and vapors, and gases (3, 4, 8).

Classification

When designing life safety for the industrial facility, the classification of the occupancy must be determined first. The exiting design requirements in the *Life Safety Code* are based on this classification (6). These classifications are dependent on the burning characteristics of the contents of the building and not on the quantity of combustible materials or type of construction. The contents are evaluated to determine the rate of fire spread, products of combustion development, and other factors that affect the time available for evacuation of the facility. Hazard classifications are general industrial, special-purpose industrial, and high-hazard industrial (3, 4, 6, 8).

General Industrial

General industrial occupancies are ordinary and low-hazard manufacturing operations (6). Examples include textile mills, automobile plants, steel mills, and clothing manufacturing.

Special-Purpose Industrial

Special-purpose industrial occupancies are ordinary and low-hazard manufacturing operations in facilities designed for particular types of operations with a low employee population density (6). The areas are normally occupied by machinery or equipment. Since there is usually minimal life safety hazard, because employees are restricted to a very small area, exits should be designed for the total number of employees in the facility. An example might be a paper mill with large machines covering a large area where process control is handled out of a small control room (3, 4, 6, 8).

High-Hazard Industrial

High-hazard industrial occupancies include facilities with high-hazard materials, processes, or contents (6). Examples include paint and chemical plants, explosives manufacturing, grain-handling operations, or flammable liquids handling and processing. Incidental high-hazard operations (paint spray booth) or flammable liquid storage (storage room) would not cause the entire facility to be high hazard. For these cases, special precautions might be taken in the high-hazard areas or only the affected area might be designated as high hazard (3, 4, 8).

Exiting

Reference should be made to the *Life Safety Code* for detailed requirements on exiting and life safety features (6). Also, state and local building codes should be referred to for any requirements peculiar to the local area. Basically, the requirements in the *Life Safety Code* for means of egress in general industrial occupancies include many of the features required in any structure (6). The travel distance to an exit is 100 ft except that with a complete automatic sprinkler system, the distance is increased to 150 ft. In most large industrial complexes, this travel distance is hard to maintain due to the vast open areas in many plants. In some cases, exit tunnels, overhead passageways, or travel through fire walls with horizontal exits may be necessary. In most cases, though, it is not practical or economical to build exit tunnels or overhead passageways. In these cases, travel distances of 400 ft may be permitted if they are approved by local fire officials and meet the following criteria (3, 4, 6, 8):

Limit to one-story buildings.

Limit interior finish to Class A or B (flame spread less than 75 and smoke developed less than 450 in accordance with NFPA Std. 255, *Method of Test of Surface Burning Characteristics of Building Materials* (6).

Provide emergency lighting.

Provide automatic sprinkler or other automatic fire extinguishing system that is supervised for malfunctions, closed valves, and water flow.

Provide smoke and heat venting designed so that employees will not be overcome by smoke and heat within 6 ft of the floor before reaching exits.

For high-hazard industrial facilities, the travel distance to an exit is limited to 75 ft with no common path of travel. All high-hazard facilities are required to have at least two separate exits from each high-hazard area. In general industry facilities, a 50-ft common path of travel is allowed to the separate exits (3, 4, 8).

EMERGENCY ACTION PLAN

An emergency action plan (EAP) should be an integral part of the life safety element in a fire loss control program. Many emergencies at facilities—such as fire, explosion, bomb threats, chemical releases, and natural disasters—require that employees and people evacuate a building. History has shown that an EAP and adequate employee and occupant familiarity with a building can prevent disasters. A written EAP is recognized as the best way to plan evacuation and, in most occupational situations, is required by OSHA in 29 CFR1910.38, "Employee emergency plans and fire prevention plans" (1). Facility managers and/or loss control managers are responsible for developing an EAP and keeping it up to date. This section discusses the necessary elements and development of an EAP.

Responsibilities

Loss Control Manager

The loss control manager (or if there is no loss control manager, the facility manager) should be responsible for (1) overseeing the development, implementation, and maintenance of the overall facility EAP; (2) designating and training evacuation wardens (or fire wardens, floor safety officers, or similar designation); (3) reviewing the EAP with employees and building occupants, arranging to train new personnel, and notifying employees of plan changes; and (4) relaying applicable information to public fire department personnel, emergency organization personnel, employees, and evacuation wardens in event of emergency.

Evacuation Wardens

The evacuation wardens are normally appointed by the loss control manager to ensure evacuation, together with other emergency duties. The evacuation wardens may also be members of the emergency organization. They should be trained in the complete facility layout and the various evacuation routes. They should be responsible for (1) checking for complete evacuation of their designated area and notifying the public fire department and emergency organization of missing persons or location of the fire and of any trapped persons; (2) performing emergency operations as needed and detailed in the EAP, such as closing windows and doors or turning off certain electrical appliances; (3) reporting any malfunctioning alarms; and (4) assisting any disabled persons with emergency evacuation procedures.

Elements of Emergency Action Plan

The EAPs are specific to each building and should be coordinated among the various departments by the loss control manager. The following elements should be included in an EAP:

1. Preferred means of reporting fires and other emergencies
2. Emergency evacuation procedures and emergency evacuation route assignments.
3. Procedures to be followed by employees who remain to operate critical plant operations before they evacuate
4. Procedures to account for all building occupants after emergency evacuation has been completed
5. Rescue and medical duties for those employees who are to perform them
6. Names or regular job titles of persons or departments who can be contacted for further information or explanation of duties under the plan

Element 1

The EAP should direct employees to activate the nearest alarm box or other alerting mechanism in the event of a fire or should state that it may be necessary to activate additional boxes or shout the alarm if people are still in the building and the alarm has stopped sounding or if the alarm does not sound. Any unusual alarm notification procedures should be explained. Examples might include areas that have no audible alarm systems, noisy areas, areas with deaf employees, and places of assembly where a public address (PA) system should be used. The plan should instruct any person discovering a fire to notify the public fire department and the emergency organization stating the location of the fire. The plan should direct employees and occupants to call security and the emergency organization to report all other emergencies. It should direct people to give their name, location of the building, and nature of emergency and should advise people to mention specific features of the building, such as toxic substances, critical operations, disabled persons, etc.

Element 2

All employees and occupants should know where the primary and alternate exits are located and be familiar with the various evacuation routes available. Primary and alternate evacuation routes and exit locations should be described in the EAP. If a floor plan is posted, it should display the various evacuation routes and exits. The EAP should direct people *not* to use elevators as an evacuation route in the event of fire. In buildings where departments share mutual areas and/or evacuation routes, the department EAPs should be coordinated. It may be necessary to combine the department EAPs into an overall building EAP. Personnel who are regularly transient because of the nature of their jobs (e.g., facilities maintenance personnel) should be trained in general evacuation procedures.

Element 3

Critical operation shutdown is dependent on the building use. It may be as complicated as dealing with pressure vessels or fuel supply or as simple as turning off bunsen burners, hot plates, or electrical equipment. Persons assigned to these duties should be either facility management and/or evacuation wardens or those who cannot leave immediately because of danger to others if the area is abandoned, such as chemists who are conducting potential runaway reactions. These responsible people should be identified in the EAP. Ventilation system shutdown in some facilities will be controlled by automatic protection systems or at a central control station. If facility management or evacuation wardens are required to perform necessary shutdown operations, this should be identified in the EAP.

Element 4

The EAP should direct groups working together in the same area to congregate in a prearranged location identified in the EAP. It is a good idea to have an alternate location for inclement weather, if an outside location, or an alternate safe refuge area. A department organization list should be developed to provide a roster of personnel to ensure that everyone has evacuated.

Element 5

Those personnel who will perform rescue and medical duties should be identified in the EAP.

Element 6

An EAP organization list should contain the names of employees, managers, or other personnel and their job titles, positions, and relative EAP collateral duties. The EAP should include appropriate numbers to call for emergency: fire, police, medical rescue, water service, fire protection systems, utilities, etc.

Once established, the plan should be reviewed by each employee and occupant covered by the plan to ensure that they know what is expected of them in all emergency possibilities that have been planned, to ensure their safety. A well-designed EAP will anticipate possible impediments to quick and complete evacuation and will provide solutions and options for these problems. Management should make every effort to develop an EAP that includes necessary contingency planning.

EVACUATION DRILLS

Just providing exits and an emergency action plan is not enough to ensure life safety during a fire. Evacuation drills are essential so that employees and occupants of the building will know how to make an efficient and orderly evacuation.

The frequency of the drills should be determined by the amount and type of hazardous operations present in the facility and by the complexity of shutdown or evacuation procedures (3, 4). All elements of the emergency action plan should be practiced during the drill.

After each drill, the loss control manager, facility managers, and evacuation wardens should discuss and evaluate the drill to fine-tune the emergency action plan and solve any problems that may have occurred.

REFERENCES

1. *General Industry Standards—29 CFR 1910*, Occupational Safety and Health Administration, Washington, D.C., 1988.
2. *Hazardous Chemical Reporting: Community Right-to-Know—40 CFR370*, Environmental Protection Agency, Washington, D.C., 1988.
3. Cote, Arthur E., ed., *Fire Protection Handbook*, 16th Edition, National Fire Protection Association, Quincy, Mass., 1986.
4. Coté, Ron, "Life Safety in Industrial Occupancies," *Industrial Fire Hazards Handbook*, 2nd Edition, Jim L. Linville, ed., National Fire Protection Association, Quincy, Mass., 1984.
5. Lerup, Lars, Cronwrath, David, and Liu, John Koh Chiang, *Learning From Fire: A Fire Protection Primer For Architects*, U.S. Fire Administration, Washington, D.C., 1977.
6. *National Fire Codes*, National Fire Protection Association, Quincy, Mass., 1989.
7. Bochnak, Peter M. and Moll, Michael B., "Relationships Between Worker Casualties, Other Fire Loss Indicators, and Fire Protection Strategies" (internal report), U.S. Department of Health and Human Services, National Institute for Occupational Safety and Health, Morgantown, W.Va., 1984.
8. Lathrop, James K., ed., *Life Safety Code Handbook*, 4th Edition, National Fire Protection Association, Quincy, Mass., 1988.

5
Emergency Organization

One of the major elements of a fire loss control program is the emergency organization. Every facility, large or small, should have an organization to react to fires and other emergencies when they occur. This organization could range from two or three people to a completely equipped plant fire department. The organization should fit the facility. The loss control manager should consider the size of the facility, potential fire emergencies, potential exposure fires, and the availability of assistance from a public fire department. One of the important factors to consider is access to the facility by the public fire department (1–5).

The loss control manager or fire chief should analyze each department for its hazards and the probable forms an emergency would take. From this, an estimate can be made on the protection needs and the personnel needed in the emergency organization. The loss control manager should rely heavily on the department supervisors, engineers, and process developers or installers and the insurance loss prevention consultants to learn more about the hazards (1–5).

A large facility located in a rural area will need more people in its organization than will a similar sized facility in a city. The loss control manager or facility management should establish a cooperative relationship with the public fire department and should agree on how the emergencies will be handled.

Planning for the efficient handling of an emergency starts by providing some specific functions that must be performed. These apply to a range of occupancies and are basic to fire safety at the facility (2, 6). Your facility will be prepared when:

The emergency will be handled under competent control.
The sprinkler system will operate, valves will be open.
The fire pumps will operate.
The proper fire-fighting action will be taken.
The electric power and any process piping will be shut off when necessary.
Sprinkler valve operations are under strict control.
Salvage operations are started immediately.
Automatic protection (sprinklers, alarm systems, etc.) will be restored as soon as possible.
All the above functions are provided on a 24-hour basis.

The emergency organization, also known as a private emergency organization or fire brigade, should have personnel for each operating shift assigned to specific duties. Below are some primary assignments for a typical industrial emergency organization. This type of organization can be tailored to small or large businesses and various occupancies by combining or eliminating some of the assignments (2, 5, 6).

PRIMARY ASSIGNMENTS

Fire Chief

This is a very important position to fill. The fire chief, also known as chief of plant protection, should have proven leadership qualities and be able to inspire coworkers to accept and perform their duties in the emergency organization. Think of the organization as a volunteer fire department. The coworkers have their normal jobs to perform. They need to be made to feel a part of the emergency organization and that their contributions are important. The chief should have administrative and supervisory abilities (2).

The chief should

Be familiar with the entire facility and with its special hazards.
Know the automatic sprinkler system, layout, water supplies, etc.
Know the other automatic protection systems.
Be familiar with salvage and rescue operations.
Be trained and experienced in fire fighting.

Some of the chief's duties would include

Evaluation of the fire-fighting equipment.
Establishing plans of action for possible fire situations.
Review of the organization roster to maintain the roster of personnel.
Plan for training members of the organization.

There should be enough assistant chiefs appointed to cover each shift at the facility. Provisions should be made to cover during vacation, sick leave, or other periods of absence. The chief's authority must be clearly defined by the loss control manager, since the chief must direct all fire-fighting or emergency action until the public fire department arrives.

Sprinkler Valve Operator (Where Applicable)

Before determining how many sprinkler valve operators will be necessary, the loss control manager may have to look at the plans of the facility and its protective system, the number of valves that may be involved in one fire, the number of valves that one person can supervise, and whether valves are grouped together or scattered (2, 6).
Once assigned, the sprinkler valve operator should

Know the operation and location of each sprinkler valve.
Make sure that the valve stays open during a fire emergency.
Stay near the valves during the emergency.
Close the valve only on orders of the emergency organization chief or public fire chief.
Stand by after the fire, to reopen valves if necessary, until the sprinkler system is restored.

This job is usually assigned to a plumber, but can be assigned to any dependable person with the proper training. It may be practical to assign a person whose work area is near the valve. Remember, all shifts should be covered.

Fire Pump Operator (Where Applicable)

As with the sprinkler valve operators, the number of fire pump operators depends on the number, location, and grouping of the pumps (2, 6).
The fire pump operator should

Check the automatic starting pump or start the manually starting pump.
Keep the pump in operation until told to shut the pump off.

Fire Squad

Determining the size of the squad is not an easy task. The size of the facility, hazards, the pubic fire department, and other variables play a role in determining the size. It may range from one person to a fully equipped plant fire department. The squad should be large enough to bring the proper equipment quickly to the fire scene and to ensure an adequate response to control the fire, bearing in mind that the public fire department (if available) is on its way. The squad may also be accountable for rescue and cleanup operations.

The squad should be located such that it can act on the fire as soon as possible and to control or confine it. It may also be called upon to help monitor evacuation of personnel. For most facilities, the best situation is to have a group of trained employees in each area or department that can drop what they are doing and immediately respond to the fire situation. These people would best know the processes and hazards. This group would be organized into a squad with someone appointed as a leader. In most cases, this person will be the department supervisor. His or her help would be essential in selecting squad members, arranging training and practice, and fostering and maintaining their interest (2, 6).

Electrician and Plumber

These people should be qualified in their respective trades. They should have thorough knowledge of the systems in the facility. The electrician is responsible for the necessary shutting down of electrical equipment such as fans, ventilators, etc., that may hamper emergency action or endanger life. He or she may also be called upon for temporary lighting arrangements. The plumber is responsible for closing off lines such as steam, water, flammable gas, and flammable liquid piping that may interfere with emergency operations. He or she also would be responsible for replacing any opened sprinklers (2).

Salvage Squad

Salvage operations can start immediately after the alarm sounds. The best people for this assignment would be those most familiar with the equipment and products. Their duties include protecting equipment, cleaning up, and separating undamaged from damaged equipment or products. They can also close fire doors and elevator hatches (2).

Security Guard

Security serves as an adjunct to the emergency organization. For times when the work force is not normally at the facility (holidays, shutdown periods, weekends), guards may have to sound the alarm, check sprinkler valves and fire pumps, notify the emergency organization, and guide the public fire department to the area (2).

TRAINING

The personnel selected for the emergency organization should be trained in operating protective equipment, reporting emergencies, responding to fires, fire attack procedures, and operating suppression equipment. They should receive specialized training if they will be involved with sprinkler valve control, fire pump operation, electrical safety, and steam and gas piping control (2, 6). Training will be covered in more detail later under Chapter 12, Education and Training.

PREFIRE PLANNING

As stated earlier, it is very important that the facility management establish a cooperative relationship with the public fire department. One good way to establish this relationship is through prefire planning. Prefire planning is a systematic way for a public fire department to be come familiar with the facility with the assistance and cooperation of the facility management, in order to properly handle any fires at that facility. Normally, this planning is done for all target hazards and areas where major fires may occur for the specific purpose of planning fire attack. This differs from normal fire department inspections that are concerned with fire prevention and enforcement of fire regulations. The cooperation of management should include the emergency organization. In this way, members of the emergency organization can interact with the fire fighters to develop the prefire tactics together (2, 7).

The initial inspection by the fire department and facility management representatives should be thorough and cover the entire facility, with all pertinent details recorded. After that, inspection should be made periodically to update the initial data. Any fire hazards that are discovered on these inspections should be eliminated as soon as possible. The important goal in addition to preplanning for an emergency is the reduction or elimination of fire hazards (2, 7).

Facility and Protection System Layout

Preplanning should include the following general information (2, 7):

1. Location of sprinkler valves and areas controlled by the valves
2. Locations of pumper connections for sprinkler systems, so that arriving fire engines can pressurize operating sprinklers
3. Locations of private and public hydrants so that action plans can be established for various areas

4. Means of entry to the facility property and buildings to facilitate arrival to the fire area
5. Notification of any temporary, potentially dangerous situations, such as temporarily closed valves, frozen piping, new operations, etc.
6. Locations of hazardous materials, including any shutoff procedures

Interaction with Emergency Organization

Where specific hazardous processes are involved, the emergency organization should use its special knowledge and skills to help fire fighters use their skills to the best advantage in controlling the emergency. Rehearsals should be held where specific fire scenarios can be set up and the fire fighters and members of the emergency organization can practice solving the problems together (2).

REFERENCES

1. Cote, Arthur E., ed., *Fire Protection Handbook*, 16th Edition, National Fire Protection Association, Quincy, Mass., 1986.
2. *The Handbook of Property Conservation*, Factory Mutual Engineering Corporation, Norwood, Mass., 1988.
3. *National Fire Codes*, National Fire Protection Association, Quincy, Mass., 1989.
4. Tuck, C. A., Jr., ed., *Industrial Fire Brigade Training Manual*, 5th Edition, National Fire Protection Association, Boston, Mass., 1978.
5. *General Industry Standards—29 CFR 1910*, Occupational Safety and Health Administration, Washington, D.C., 1988.
6. Schontag, Peter K., "Industrial Fire Risk Management," *Industrial Fire Hazards Handbook*, 2nd Edition, Jim L. Linville, ed., National Fire Protection Association, Quincy, Mass., 1984.
7. Kimball, Warren Y., *Fire Attack—1 Command Decisions and Company Operations*, National Fire Protection Association, Boston, Mass., 1966.

6
Fire Hazards of Materials and Their Control

The intent of this chapter is to indicate classes of materials that may be considered hazardous due to their properties; to point out how these substances may be safely handled and controlled; and, finally, to discuss suitable extinguishing agents that can be used to control fires should they occur. By necessity, certain individual substances presenting unusually severe hazards are also included. The discussion will cover flammable liquids, gases, and solids, including the dust explosion problem created by certain finely divided organic and inorganic substances.

FLAMMABLE AND COMBUSTIBLE LIQUIDS

In Chapter 2, Characteristics and Behavior of Fire, certain properties of flammable and combustible liquids, including flash and fire points, upper and lower explosive limits, and the flammable or explosive range, were defined and their importance to the fire problem explained. Two additional properties of flammable liquids, not previously mentioned, are also very important in loss control. These two properties, specific gravity and vapor density, determine whether water can be used for extinguishing and where ventilation intakes should be located to control emission of vapor.

Specific gravity is defined as the relative weight of a liquid compared with water. Most flammable liquids have a specific gravity of less than 1 (with water having a specific gravity of 1), which simple means that they are lighter than water and will float on the surface where they can continue to burn freely and spread over a considerable area. Those flammable and combustible liquids that have specific gravities greater than 1, indicating that they are heavier than water, can be extinguished by water, which will cover the surface of the flammable liquid and exclude the air (oxygen) necessary for combustion. Therefore, it is quite obvious that the specific gravity is extremely important in the determination of whether water can be used as an extinguishing media for a particular liquid.

Vapor density may be defined as the relative weight of the vapors of a liquid as compared with air. This particular property is extremely important in enabling properly designed ventilation to remove hazardous vapors from flammable liquids to a safe point. The vapor densities of all flammable liquids are greater than 1, meaning that the vapors will settle to low points and may travel along floors over considerable distances and accumulate in pits, trenches, and other subsurface locations. This requires ventilation pickup points at low levels rather than overhead. High-level pickup is consequently of limited value in reducing vapor concentrations to an acceptable level, which is generally considered to be 25% of the LEL.

Another property of flammable and combustible liquids of importance in fire protection is *miscibility* with water. This means the ability of a liquid material to mix with water. Flammable liquids that are miscible with water can have their flash and fire points raised by the addition of water; moreover, with water, they become conductors of electricity, thereby helping to dissipate static charges. Certain solvents, generally other than hydrocarbons, having high degrees of miscibility with water will deteriorate regular fire-fighting foams. These liquids, termed *water-soluble* solvents or polar solvents, include the alcohols, ketones, and esters and require special foams, generally referred to as *alcohol* foams, for effecting extinguishment.

Properties of flammable and combustible liquids and other materials are listed in various publications, including the NFPA *Fire Protection Handbook* (1) and the FM *Loss Prevention Data Books* (2). The NFPA also publishes *Flash Point Index of Trade Name Liquids*, NFPA 325A (3), and *The Fire Hazard Properties of Flammable Liquids, Gases and Volatile Solvents*, NFPA 325M (4).

Classification

To distinguish between gases and solids, flammable liquids are classified as liquids with vapor pressures of 40 psi or less and with a fluidity greater than that of 300 penetration asphalt. Table 6.1 lists the flammable and combustible liquid classification system that is defined in 29 CFR1910.106, *Flammable and Combustible Liquids* (5) and the *Standard on Basic Classification of Flammable and Combustible Liquids*, NFPA 321 (4).

It is important to recognize and properly classify the flammable liquids that are handled. The federal regulations 29 CFR 1910.106 (5), the *Flammable and*

Table 6.1   Flammable and Combustible Liquid Classification

Type	Definition	Examples
Flammable		
Class IA	FP < 73°F (23°C) BP < 100°F (38°C)	Pentane, ethyl ether
Class IB	FP < 73°F (23°C) BP ≥ 100°F (38°C)	Alcohol, acetone
Class IC	73°F (23°C) ≤ FP < 100°F (38°C)	Turpentine, xylene
Combustible		
Class II	100°F (38°C) ≤ FP < 140°F (60°C)	Kerosene, mineral spirits
Class III A	140°F (60°C) ≤ FP < 200°F (93°C)	Pine oil, nitrobenzene
Class III B	FP ≥ 200°F (93°C)	Vegetable oils, ethylene glycol

Note: FP = Flash point
      BP = Boiling point

Combustible Liquids Code, NFPA 30 (4), and many Fire Prevention Codes adopted by municipalities across the nation impose limitations on storage and sizing of containers for these liquids, with more stringent restrictions for the more hazardous materials.

Fire Protection Considerations

Fires involving flammable and combustible liquids cause extensive property damage, and the handling and storage of these materials are a threat to personnel safety. Vapors from flammable liquids can permeate large enclosures; can be readily ignited with relatively low-energy ignition sources; and, when ignited, have very high energy releases that make fires difficult to extinguish. In turning our attention to the control of these materials, consideration should be given to basic principles for controlling not only flammable liquids, but other hazardous materials as well, as outlined below:

1. Segregate the hazard.
2. Confine or enclose the hazard.
3. Ventilate to prevent explosive mixtures.
4. Install explosion venting where needed.
5. Eliminate sources of ignition.

6. Educate those involved as to the hazards and safeguards.
7. Provide adequate fire protection.

Segregating the Hazards

It is axiomatic when planning for control or protection of hazardous materials to think in terms of isolating these materials or hazards from other operations, from areas of high value concentrations, from vital plant facilities such as power supplies, and from locations that might possibly involve life safety, such as means of egress. Any discussion of isolation of a hazard must invariably consider, as an initial approach, the separation of the hazard by distance. The first question, therefore, must be, "Can this hazard, whether it is a storage or processing hazard, be in an open area away from other facilities, or can it be in a detached structure?"

While product transfer involving pump and piping systems is most desirable from the standpoint of good loss control practice, the corporate planner or plant facilities manager must consider the higher costs involved—in all probability, additional structural costs, and certainly the coverage of greater land area. For the more volatile and dangerous operations, the need for greater fire safety will frequently be an offsetting factor to economic considerations.

Often, space limitations will preclude the separation of a hazardous operation by distance, presenting the next consideration, in descending order of priority, of physically separating the hazard by providing enclosures with a fire resistance rating of sufficient magnitude to segregate or confine the fire in the event of ignition.

The construction of walls or other structural elements imposes heavy economic burdens, suggesting that more reliance be placed upon protective systems. Fire resistance ratings for more common flammable liquid operations will be discussed in other sections of this chapter.

Hazard Confinement

In discussing enclosures to confine flammable liquid hazards, special consideration must be given to openings in fire or fire-barrier walls that prevent the escape of flammable liquids from the enclosure to surrounding areas. This can be done by providing grated drainage trenches across all openings with drains to a safe location, by providing curbing or ramping at the doors a minimum of 4 in. in height, or by having the floor of the flammable liquid area at a lower elevation than surrounding floor area.

Any consideration of the control of flammable liquids must include the safe handling of these materials. Wherever possible, liquids should be confined, whether handled in small amounts as reagents or in tank-car quantities for processing. Safety containers are available in various sizes, and their regular usage is strongly recommended (Fig. 6.1).

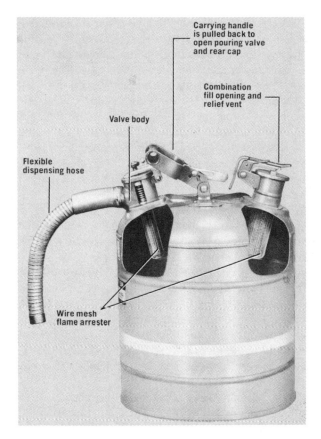

Figure 6.1   Safety container used for handling flammable liquids (Courtesy of Justrite Co.)

In the preceding section the desirability of isolating hazards was pointed out, as was the desirability of confining any flammable liquid spills to the hazard area. It is well recognized that the safest method of handling flammable liquids is by enclosing piping systems properly designed so that, in case of physical damage to the piping, causing leakage, the liquid will not continue to flow by gravity, by a pressurizing agent, or by siphoning. Pumps should be of a positive-displacement design, since this type provides a safe shutoff, preventing siphoning when not in operation. This is a disadvantage when using centrifugal pumps in flammable liquid handling. When selecting piping systems, consideration should be given to emergency shutdown of the flammable liquid, which will include designing for both human and mechanical failure. The use of a pumping system with positive-displacement pumps is the preferable method with this type of system, provided a relief valve of adequate capacity is used to prevent overpressurizing the piping. For the same reason that gravity flow is undesirable, pumps should also be arranged to take suction under lift rather than under a constant head. The use of hydraulic displacement or compressed gas displacement also presents the disadvantage of the inability to stop the flow of the flammable liquid in the event of an upset condition. Wherever possible, pumps for handling flammable liquids should be located outdoors away from main buildings or valuable installations. As a less desirable alternative, pumps should be placed in detached pump houses or cutoff rooms with construction of a noncombustible type. Pumps are a frequent source of leaks and, where possible, should be protected by water-spray or closed-head sprinklers for control purposes.

Materials used in piping should be resistant to the corrosive properties of the liquid handled, and, in addition, the material should be resistant to heat and mechanical damage. The use of plastic piping with relatively low melting temperatures should be avoided in handling flammable liquids if escape of flammable liquids under fire conditions is to be prevented. To ensure compatibility with liquids handled, gaskets, flanges, and joint compounds should be selected with considerable care to avoid weak points in a piping and pump system. The use of adequate strength piping materials is essential to withstand maximum service pressures. Generally, Schedule 40 pipe should be used for up to 120-psi service pressure, and Schedule 80 pipe should be used for service pressures between 125 and 300 psi. Fittings should match the pressure rating of the piping. Cast iron pipe should not be used.

The location of the piping is an important consideration that should not be overlooked. Piping of flammable liquids should not be located in main building areas if at all possible. It is desirable to run piping outdoors either underground or attached to the side of a building, preferably beneath any possible wall openings and protected from physical damage. Considerable care is essential in planning the location of interior piping systems that traverse main or vital areas. Such installations should be located under automatic sprinkler system protection. Placing valves on all piping systems at safe and accessible locations will permit manual control of flow. Automatic or manual means for stopping flow under emergency situations are often desirable. These include the use of so-called dead-man controls requiring constant attention of the operator, heat-actuated valves held open by fusible links, pressure controls designed to shut down on loss of pressure, and excess-flow-type valves.

Special considerations are required for bulk flammable liquid storage in tanks because of the potential for large losses. These include the location of tanks with reference to buildings and adjacent properties; the confinement of any escaping liquids or their diversion away from high-valued facilities; and provisions for normal venting of tanks, for emergency venting under fire conditions, and for extinguishment.

Tanks are usually constructed of steel, but other materials are permitted by code, particularly if the liquid stored is not compatible with the metal, resulting in either contamination of the product or corrosion of the steel. However, it is important that tanks located above ground or within buildings be of noncombustible construction. Glass fiber-reinforced plastic tanks are listed by approval agencies, such as Factory Mutual (6) for underground storage with the provision that the metal fill or suction line should extend nearly to the bottom of the tank to dissipate possible static charges that may accumulate during filling. Tanks may have combustible or noncombustible linings.

While tanks are classified according to pressure as either atmospheric pressure (designed for pressures from atmospheric through 0.5 psig), low pressure (for pressures above 0.5 psig but not more than 15 psig), or pressure vessels, the fire protection engineer is normally involved in atmospheric or low-pressure types. Large quantities of flammable or combustible liquids are stored in vertical tanks of different types, with cylindrical tanks most generally used either horizontally on supports or installed vertically.

The design of horizontal tank supports requires attention to prevent collapse under fire conditions. Tank supports should be of concrete, masonry, or protected steel. A minimum of 2-hr fire resistance protection for steel supports or exposed piling should be provided, except that approved water-spray protection or its equivalent is an acceptable alternative. A 2-in. coating of concrete (on wire mesh) protecting the steel will provide this degree of resistance. A mastic material is also available, but its intumescent capabilities may be affected by weathering. Firm foundations are necessary for tanks as well as for any other structure. Adequate supports will prevent uneven settlement, thereby avoiding the possibility of undue stressing of pipe connections. Concrete ring walls are installed for support of vertical tanks where poor soil conditions are encountered. Water spray systems are frequently used in lieu of physical coatings or coverings for protection of tank or other vessel supports and are considered an excellent alternative to this protection.

Outside storage tanks should be located to avoid exposure to other facilities or property lines, with site location involving separation by distance, confinement of the liquid in the event of tank rupture, or diversion of any possible liquid flow away from important facilities. Tables specifying minimum distance from property lines, from any public way, or from nearest important buildings are specified in 29 CFR 1910.106 (5) and the *Flammable and Combustible Liquids Code*, NFPA 30 (4). It should be noted that these tables specify minimum distances, and in many cases, additional separation by distance should be provided in the interest of good loss prevention practices. These tables state that for storage tanks having capacities up to 30,000 gal, the minimum distance from the nearest important building shall be 5 ft. This obviously is inadequate in the case of a highly volatile material exposing a building having either combustible wall construction or unprotected wall openings. The minimum distance between adjacent flammable or combustible liquids as required by code is 3 ft, with this shell-to-shell spacing also listed, with a minimum separation between a flammable or combustible liquid storage tank and a liquefied petroleum gas container required to be 20 ft.

Tanks of an above-ground tank farm should be located on ground draining away from important facilities, which may require diverting walls. Where this is possible, a slope of not less than 1% away from the tank toward the drainage system should be provided, with termination of the drainage system in a safely located impounding area having a capacity of not less than the capacity of the largest tank served. In lieu of drainage for protection of exposed property or waterways, confinement of liquid around the tank should be by means of a dike having a volumetric capacity of not less than the greatest amount of liquid that can be releases from the largest tank within the diked area (assuming a full tank). Dikes may be of earth, steel, concrete, or solid masonry construction and of a liquid-tight design adequate to withstand the full hydrostatic head anticipated. To prevent entrapment of personnel, dike heights should be restricted to an average height of 6 ft above interior grade.

In a dike area containing two or more tanks, subdivision—preferably by drainage channels or, as a minimum, by intermediate curbs—should be provided to prevent spills from endangering adjacent tanks within the diked area. Intermediate curbs should be a minimum of 18 in. in height.

Venting

All tanks should provide for the escape of air during filling operations, for the influx of air at times of flow from the tank, and for pressure changes created by variations in temperature. Provision of venting depends upon filling and emptying pumping rates, the characteristics of the liquid, and the size and design of the tank. Normal venting to prevent the development of vacuum or pressure, to avoid distorting the tank or exceeding the design pressure, should be in accordance with the requirements of the *American Petroleum Institute's Venting Atmospheric and Low-Pressure Storage Tanks*, API Standard 2000 (7).

Venting devices for the more volatile flammable liquids are normally closed, except when tanks are under vacuum and pressure, with the main objective of conserving the contents by preventing excessive loss through evaporation. Venting devices for Class I liquids should be provided with flame arresters, except where liquid properties are such that condensation or crystallization and freezing of moisture in winter make the use of flame arresters impractical. The object of flame arresters is to prevent flashback of fire to tanks where the vapor mixture may be in the explosive range. Arresters dissipate heat by means of banks of par-

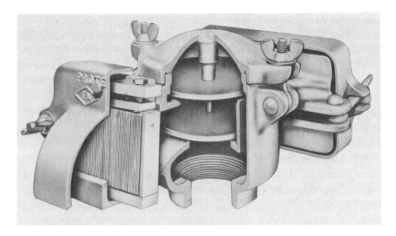

Figure 6.2 A typical conservation vent used on storage tanks containing flammable liquids. (Courtesy of Protectoseal Co.)

allel metal plates or tubing having large surface areas of metal, with the heat adsorption lowering the temperature of the vapor below the ignition point. For small vents, screens of 40 mesh will ordinarily prevent the passage of flame through the opening, but it must be recognized that these vents are subject to clogging and physical damage. Course screens, provided on many vents, including conservation vents, cannot provide any flame-arresting capability. Conservation vents, which are used with most volatile materials and are closed except under vacuum and pressure conditions, can also be provided with flame arresters. A typical conservation vent is shown in Figure 6.2.

In addition to venting for normal in-breathing and out-breathing, provisions for relieving excessive internal pressure caused by exposure fires should be provided. This relieving of excess internal pressure, termed *emergency relief venting*, may be in the form of a weak roof-to-shell seam or other approved pressure relieving construction, such as the use of a type of cover or vent weighted to stay closed except under the force of internal tank pressure. Standards 29 CFR 1910.106 (5) and NFPA 30 (4) should be consulted for calculation of an adequate emergency relief venting capacity.

Where vent pipes from tanks handling Class I liquids are adjacent to buildings, caution in the termination of the vent pipe must be exercised so that vapors will be released to a safe location outside the building, not less than 12 ft above adjacent ground level. Vent pipes terminating under eaves will trap flammable vapors in these areas and, therefore, this condition should be avoided. Also, vents should not terminate within 5 ft of any building opening.

Outside buried tanks offer the safest method of handling flammable liquids, but also require some judgment in location, particularly when in the vicinity of buildings containing basements. Standards 29 CFR 1910.106 (5) and NFPA 30 (4) permit buried tanks handling Class I liquids to be located not less than 1 ft from any basement or pit or 3 ft from any property line that may be built upon.

Wherever practical, tanks should not be located inside buildings, although tanks often are located inside buildings in industrial plants, processing plants, refineries, and chemical plants. Venting for interior tanks to a *safe* outside termination should be the same as that provided for exterior tanks, with the exception that emergency venting by use of weak roof seams should not be permitted. Every tank opening, excluding vents, should be provided with a valve located as close as practical to the shell of the tank. This provision should also be followed in tank openings on outside tanks where liquid may escape. To prevent possible collapse of tanks, protection of supports should be provided as described earlier in this chapter.

Storage Areas

While it is recognized that the safest method of handling flammable liquids is by an enclosed piping system, it is understood that there are many situations that will require the use of containers for storage and handling of flammable liquids. This involves drums, portable tanks, and small containers.

Drums and containers are normally considered to be 60 gal or less. Storage of these containers should preferably be in detached buildings located a minimum of 50 ft from the nearest building. But this may not always be practical. If containers must be kept inside, they should preferably be in cutoff rooms or attached buildings. The NFPA *Industrial Fire Hazards Handbook* (8) lists "some storage arrangements that follow the precepts of protection by good separation distances or construction." These are listed below in order of their preference and coded to Figure 6.3 (8).

1. Outside yard areas at least 50 ft from buildings, structures, combustible storage, or property lines (separation distance).
2. Detached buildings, lightweight construction, at least 50-ft separation distance.
3. Detached buildings, lightweight construction, automatic fire detection and suppression, 10- to 50-ft separation distance.
4. Attached one–story addition, standard 4-hr fire-wall separation.
5. Attached one-story addition, automatic fire detection and suppression, 2-hr fire partition separation.
6. Inside building in a corner with interior walls and ceiling having at least 3-hr fire resistance rating and pressure resistance rating of 100 lb/ft^2 and exterior walls, lightweight, pressure-relieving construction.

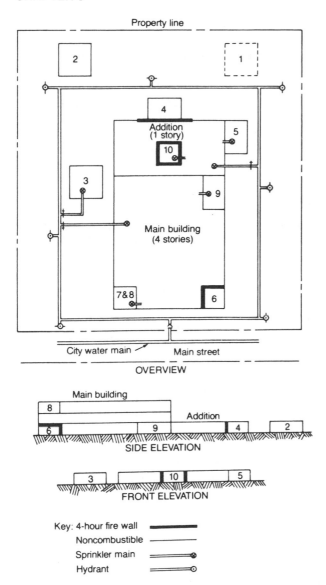

Figure 6.3 Recommended storage locations numbered in order of preference. (Reprinted with permission from the *Industrial Fire Hazards Handbook* 2nd Edition, Copyright 1984, National Fire Protection Association, Quincy, Mass.)

7. Same as item 6, with automatic fire detection and suppression and interior walls and ceiling with 2-hr fire resistance rating.
8. Same as item 7, located in upper stories with drainage and liquid-tight floors.
9. Inside building, at or above grade level, with interior walls and ceiling with 2-hr fire resistance rating, exterior wall of lightweight, pressure-relieving construction, and automatic fire detection and suppression.
10. Totally within building, at or above grade level, automatic detection and suppression and enclosure walls and ceiling with 2-hr fire resistance rating.
11. Unprotected areas in buildings are the least desirable and should be avoided.

The features of inside flammable liquid storage rooms are well defined in NFPA 30 (4) and should be adhered to in the interests of good loss control practices. Up to 5000 gal are permitted in a maximum area of 500 ft^2, with a maximum of 10 gal allowed per square foot of floor area. The enclosure for this area is required to provide a minimum of 2 hr of fire resistance, protected by either closed-head sprinklers, water spray, carbon dioxide, Halon, or dry-chemical systems.

This same room, without any of the protection systems described above, may contain up to 2000 gal of flammable liquid with a concentration of 4 gal/ft^2 of floor area. Rooms of 1-hr fire resistance rating are restricted to a total area of 150 ft^2 with a maximum flammable liquid capacity of 750 gal if protected by one of the automatic systems mentioned above, or 300 gal if unprotected.

In addition to the fire resistance rating of the inside storage enclosure, openings to the other rooms or buildings should be provided with noncombustible, liquid-tight, raised sills or ramps, a minimum of 4 in. in height, or the floor in the storage area should be at least 4 in. below the surrounding floor. A permissible alternative to the sill or ramp is an open-grated trench inside the room that drains to a safe location. An important point to keep in mind is that drains must be adequately sized, along with any salvage tank that may be provided. The tank will not only drain off a flammable liquid spill, but also will drain water from automatic sprinklers and hand hoses. Inlets must also be cleaned frequently to prevent clogging.

All inside storage rooms should be equipped with either gravity or mechanical exhaust ventilating systems. Where dispensing is done in the room, only mechanical ventilation should be used to provide a minimum of 1 ft^3/min per square foot of floor area with a minimum of 150 ft^3/min. Any ventilating system for flammable liquid vapors should be provided with low-level pickup and, in addition, ventilation should be interlocked with the room lighting.

If flammable liquid storage rooms are used merely for storage, electrical wiring and equipment of Class I, Division 2 are adequate. However, if dispensing is done from the room, Class I, Division 1 provisions should apply. Figure 6.4 shows an example of a dispensing area.

Containers of over 30-gal capacity should not be stacked upon one another. Dispensing from drums or portable tanks should be by approved pumps or by self-closing faucet. Hand-operated drum pumps, consisting of a pumping unit equipped with discharge fittings and a means for mounting in a bung opening of a drum, are available. Pumping units also provide for drum venting. In addition, only approved self-closing faucets should be used for dispensing. This will overcome the obvious disadvantage of conventional-type valves being left open. Plastic faucets, found frequently in industry, should never be used. Safety bungs should also be provided when dispensing from drums. These bung vents are specially designed to relieve internal pressure at 5 to 15 psi and are equipped with a flame arrester of the screen or plate type. Under fire conditions, safety bungs will permit the release of pressure and burning at the bung, preventing container rupture and the resulting release of flammable liquid.

Maximum allowable sizes of containers and portable tanks are listed in NFPA 30 (4) and 29 CFR 1910.106 (5).

Relatively small quantities of flammable or combustible liquids not requiring a flammable liquids storage vault should be stored in metal cabinets constructed of double-walled, minimum 18-gauge sheet iron with 1-1/2-in. air space, having riveted or welded joints, with a door provided with a 3-point latch arrangement and a door sill at least 2 in. above the bottom of the cabinet. These cabinets, constructed according to the above specifications, are available commercially. They conform to requirements of NFPA 30 (4) and 29 CFR 1910.106 (5) and may be used for storage of not more than 60 gal of flammable liquids or 120 gal of combustible liquids.

Flammable liquids not stored in a room especially constructed for this purpose or in a flammable liquids cabinet should be handled only in minimal quantities and then only in approved safety containers. Such containers are designed to provide vacuum and pressure relief and are provided with self-closing covers (Figure 6.1). Stainless-steel types are available for liquids that may be contaminated by steel. These containers are available to a maximum capacity of 5 gal.

Storage of flammable liquids in drums on racks presents a severe challenge to a fire protection system because the possibility of rupture of drums from fire and the spread of flammable liquid, exposing other drums. Racks containing more than two levels of drum storage should be provided with in-rack sprinkler protection. The design of sprinkler protection for flammable liquid areas requires careful consideration, with provision for an adequate density of water over larger areas because of the rapid and high rate of energy release from fires of this kind.

In addition to safety cans, where it is necessary to moisten a material with a volatile flammable liquid, approved plunger cans are available that force liquid

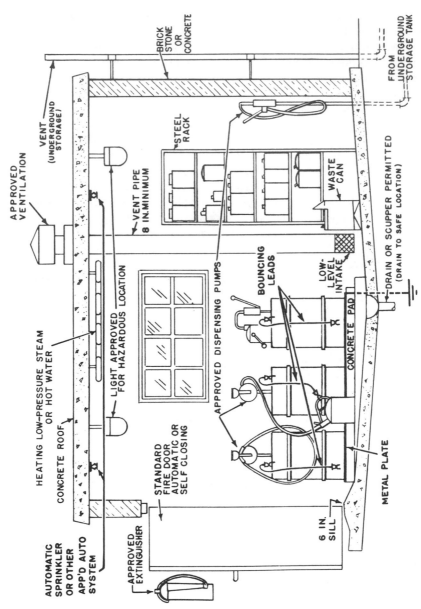

Figure 6.4 Example of a flammable liquid dispensing area. (Reprinted with permission from the National Safety Council. *Accident Prevention Manual for Industrial Operations*, 9th Edition. National Safety Council, Chicago, 1988, p.412.)

up through a hollow plunger when the top and piston are depressed to moisten the screen in a concave receiving pan. The screen serves as an effective flame arrester. Approved small dip tanks are also available in sizes up to 2 ft^2, equipped with automatic closing covers actuated by fusible links for coating, cleaning, or dipping of small parts.

As indicated previously, extensive flammable liquid operations should be segregated by isolating or confining the hazard. The first order of priority, depending upon practicality, is to separate the area involved by distance, using as a rule of thumb a minimum of 50 ft. Certain operations of a highly hazardous nature might, by necessity, require additional separation by distance or the construction of explosion-resistant exposed walls.

The use of explosion-resistant walls is of great importance when a highly hazardous operation must be located adjoining a main building area. These walls are usually constructed of reinforced concrete to withstand pressures of about 100 lb/ft^2, although lighter walls of metal lath and plaster or metal panels are also used for this purpose. Walls of this nature are flexible, having a tendency to bow out under the pressure buildup of an explosion, as compared with the fragmenting of unit masonry-type walls such as concrete block. Unit masonry-type construction should not be used where explosions may be anticipated, unless they are reinforced.

Explosion Venting

Where explosion possibilities exist, as in the use of highly volatile flammable liquids, explosion-relief venting should be incorporated to vent explosion pressures in a safe direction. This additional feature will help an explosion-resistant wall to withstand the force of the explosion by venting pressures, thus preventing their reaching levels exceeding the design pressure.

The location of flammable liquid operations, either in or adjoining buildings, is obviously of extreme importance. All floors of multistory buildings (with perhaps the exception of top floors), basements, and interiors of one-story buildings are undesirable, since it becomes extremely difficult, if not impossible, to provide explosion-relief venting and, in some cases, adequate ventilation. A structure housing a hazardous, explosion-prone operation, for example, should not expose an existing building, upper stories, or vital utilities. Proper construction using explosion-resistant walls and explosion-relief venting and location are features to consider in protection and planning.

Explosion-relief venting with a minimum ratio of 1 ft^2 for each 50 ft^3 of room volume may be provided by explosion-relief-type windows, light plastic or metal panels, the use of necked-down, frangible-type bolts securing wall panels, etc., designed to relieve from 20 to 25 psi pressure. The actual explosion-relief-venting ratio will depend upon the hazard involved, and a ratio providing more venting than 1 ft^2 for each 50 ft^3 might be required to prevent unacceptable structural damage.

Emergency drainage should be provided in the event of flammable liquid spillage and also to remove fire protection water to a safe location. Systems that permit the discharge of liquids to adjoining properties, sewers, or waterways should be prohibited except in the case of discharging to public sewers or waterways, where such systems are equipped with traps or separators. If drainage is to a tank, it is imperative that it be sized to handle water discharged from sprinklers and hoses to prevent overflow into adjacent areas.

The design of equipment used in flammable liquid operations should include enclosed transfer systems and equipment to confine vapors. Equipment subject to internal explosion possibilities should be designed to withstand an explosion. The incorporation of explosion-relief venting to the outside wall materially reduce costs by lessening design requirements and permitting lighter construction. In addition, it should be kept in mind that tanks or vessels used for processing should be provided with fire-resistive or protected steel supports where an exposure fire of sufficient duration to cause the collapse of a vessel could occur.

All portions of process areas where flammable liquid vapors are normally present due to open containers, dispensing of these materials, etc., should be provided with ventilation as a basic prevention against the formation of flammable liquid vapor concentrations within the explosive range. Ventilation should be at a rate of not less than 1 ft^3/min per square foot of solid floor area. This may be accomplished by natural or mechanical ventilation, with discharge or exhaust to a safe location outside the building. It is recommended that, for operations involving Class I liquids and those liquids heated above or near their flash points, continuous mechanical ventilation be provided.

Controlling Ignition Sources

All ignition sources should be eliminated. These sources include, but are not limited to, open flames; smoking; cutting and welding; hot surfaces; friction of heat; static, electrical, and mechanical sparks; spontaneous ignition; and radiant heating.

Static electricity has caused numerous fires in flammable liquid operations. Static charges are produced when the liquids interface, by making and breaking contact with other materials. Typical contacts are the flow through pipes (liquid and pipe), the pouring of flammable liquid from one container to another (liquid and air), the filling of a tank by permitting free fall of the liquid within the tank (liquid and air), and mixing and agitating. If there is no way to permit their dissipation, static charges accumulate. If the accumulation is sufficient, a static spark occurs, resulting in ignition. Of course, another condition must exist, that being the existence of a vapor concentration within the flammable or explosive range. Different flammable liquids exhibit different static-generating tendencies or ability, depending upon the *resistivity* of the liquid, which may be defined as the measure of its ability to hold a charge. Tests have indicated that the higher the resistivity, the greater the ability of a liquid to hold a charge. Liquids having a resistivity higher than 10^{10} ohm-cm may accumulate a charge. If the resistivity of a liquid is less than 10^{10} ohm-cm, any charges generated will recombine without accumulating to a hazardous degree. Approximate resistivities of some flammable and combustible liquids are listed in *Recommended Practice on Static Electricity* NFPA 77 (4). This standard indicates that flow through metallic piping generates static, but experience has shown that closed-piping systems present no static hazard. Further, bonding is not needed around flexible metal piping or metallic swing joints even though lubricated, but bonding is required around joints in which the only contacting surfaces are made of nonmetallic insulating materials.

Although grounding will not be effective in all cases involving surface charges, it is recommended that all equipment containing Class I flammable liquids be electrically grounded, particularly at points of transfer of liquid from one container to another. A typical arrangement of the transfer of a flammable liquid from a drum to a safety can using the drum ground and bonding connection is shown in Figure 6.5.

To prevent the buildup of static charges in filling containers, conductive fill pipes should be provided within 6 in. of the bottom of the container, with the containers electrically grounded and bonded. *Bonding* is simply the interconnecting of the containers to prevent a potential (voltage) differential and is not required when a container is filled through a closed system.

In filling metal containers from drums, the fill spout, nozzle, or fill pipe, if conductive, should be kept continuously in contact with the edge of the filling opening to avoid the possibility of a static arc. In this situation, bonding is not completely necessary but is advisable. Grounding of containers used in transfer is often done by having a common conductive plate, which obviates the necessity of also providing a bonding wire. This common conductive support is often a conductive floor such as used in hospital operating rooms. These floors, on occasion, are also found in industrial plants where it is desirable to prevent accumulations of electrostatic charges. The floor simply serves as a means of connecting together the objects to minimize the possibility of an electrical discharge.

Grounding or bonding is normally obtained by wires or metal straps, which, to dissipate any static charges, theoretically may be of small size. However, to protect against physical damage, a minimum copper wire size of No. 6 AWG is recommended, and although either insulated or noninsulated wire is acceptable, noninsulated wire is preferable to indicate immediately any obvious physical break. Maintenance of connections for providing static grounding and bonding is often a problem, and any self-inspection program in flammable liquid areas should include this as a standard item.

Figure 6.5 The safe transfer of a flammable liquid from one container to another with ground and bond connections. (Courtesy of Justrite Co.)

Humidification, while not deemed practical in most industrial applications, is used in hospital operating rooms as a way to assist in the dissipation of any static charging, but is by no means foolproof. Generally, humidification ranges of 50% to 55% are maintained, whereas 60% to 70% of relative humidity would be required to provide reasonable assurances that static charges will be dissipated.

Ionization is another way to dissipate the static charges that are produced on nonconducting objects such as rubber or leather conveyor belts or on paper cloth in coating operations. Static charges are dissipated from these nonconductive materials when brought in contact with ionized air. Ionization is accomplished by heat, high voltage, ultraviolet light, radioactivity, or nonelectrical energized static collectors or neutralizers. This equipment is characterized by having sharply pointed construction such as tinsel and bronze or copper wire or bristles that produce an electrostatic field when an electrostatic field when an object charged with static electricity is brought near, producing a voltage on the points and causing the air to be ionized. The static charge is reduced to below the arcing point by draining off the charge or by neutralization. Electrically energized neutralizers are also used that produce a field between electrified points and a grounded member and that are located where static charges are likely to accumulate. The air is ionized by the high-voltage field, and charges are neutralized.

GASES

This section covers a discussion of gases that are commonly encountered in industry. Oxygen, even though it will not burn, has been included because it is a strong supporter of combustion. Flammable gases are similar in hazard to flammable liquids, possessing many of the properties and problems previously discussed. Indeed, many gases today are shipped in a liquefied form under low temperatures or pressure and, at a higher temperature or with release of pressure, the liquid becomes gases.

Gases do present, in some instances, considerably greater hazards than do flammable liquids because they will occupy the entire volume of an enclosure. Most are under pressure, creating the need to halt the flow of gas under emergency conditions to avoid the possibility of involvement of very large areas or volumes. Fire fighters know that burning gases under pressure must not, under any circumstances, be extinguished unless it is possible to control the gas flow.

Classification

For the general purpose of distinguishing a gas from a flammable liquid vapor, a substance that has a vapor pressure greater than 40 psia at 100°F (38°C) can be considered a gas. A gas normally means a substance that exists in the gaseous state at normal temperature and pressure (approximately 70°F—21°C—and 14.7

psia). Gases can be classified by chemical properties, physical properties, and usage (1).

According to chemical properties, gases can be classified as flammable, non-flammable, reactive, or toxic. A flammable gas is simply any gas that will burn in normal concentrations of oxygen in air. Like flammable liquid vapors, flammable gases have a flammable range and a finite ignition temperature.

A nonflammable gas will not burn no matter what the concentration in air. Gases in this group can be oxidizers that support combustion, like oxygen, or inert gases that do not support combustion, such as nitrogen, argon, or helium (1).

Reactive gases are those that react chemically with other gases or within the gas itself, normally emitting hazardous quantities of heat. For example, fluorine is highly reactive with most substances. Some gases, such as acetylene and vinyl chloride, can undergo chemical changes when subjected to heat and shock.

Toxic gases can create a serious health hazard by being poisonous or corrosive. Some examples are chlorine, hydrogen sulfide, and carbon monoxide.

Gases can also be classified by physical properties. Compressed gases exist in a gaseous state under pressure at normal temperatures in a container. The pressures can range from 25 to 3000 psig. Examples of compressed gases are oxygen, ethylene, hydrogen, and acetylene. Liquefied gases exist in a partly gaseous and partly liquid state under pressure at normal temperatures in a container. Normally, a liquefied gas is much more concentrated than is a compressed gas, usually at orders of magnitude more volume when released to the atmosphere. Examples are liquefied petroleum gas (LPG) and methylacetylene-propadiene, stabilized (MPS).

A *cryogenic* gas is a liquefied gas in a container maintained at temperatures will below normal atmospheric temperatures at low to moderate pressures (e.g., liquefied oxygen at –297.4°F–183°C–and liquefied methane at –258.7°F –161°C) (1).

Finally, gases can be classified by use. Fuel gases refer to flammable gases burned with air to produce heat. The most widely used are natural gas and liquefied petroleum gas. Industrial gases include the range of gases used in chemical processes; i.e., heat treating, chemical processing, welding and cutting, etc. Medical gases are specialized in the area of medical treatment; i.e., anesthesia, respiratory therapy, etc.

Fire Protection Considerations

As with flammable liquids in the previous section, consideration should be given to the same principles for controlling flammable gases.

Segregation

In general, for cylinders and small tanks, flammable gases should be separated by at least 20 ft. As a substitute for this distance, a noncombustible barrier (as high as the containers) of a least a 1/2-hr fire resistance rating can be used. Flammable gases should not be stored next to oxidizing gases. Combustibles should be eliminated in the gas container area. The storage rooms should be preferably of noncombustible construction (not necessary for small amounts: one or two cylinders).

Confinement

To prevent the escape of a flammable gas and the consequent forming of a flammable mixture in the air, the gas must be confined in a container or piping system of a design that is adequate to withstand pressures and temperatures to which it may be subjected, to be leak free, and to be located so as to avoid physical damage. For the most part, gases found in industry are stored and used in small containers. There are a few large-volume users that will have gases piped or contained in large storage containers and, of course, natural gas is piped to business and industrial sites. To transport gases economically, the gases are either compressed, liquefied, or cryogenic, as explained earlier. This also makes the gases more hazardous due to the high pressures and expansion rates.

Ventilation

All indoor storage areas should be ventilated. The ventilation rate can vary, depending on the hazard and arrangement of the storage. If there is strictly storage and no filling of containers involved, the ventilation rate may be low (around one air change per hour). Normally, this rate will be adequate when the concentration of the gas in the gas-air mixture does not exceed 25% of the lower flammable limit.

Most common gases found in industry are lighter than air and, as a result of this characteristic, can be more readily controlled. Liquefied petroleum gases are the exception, because these gases, such as propane and the butanes, are heavier than air. These gases will settle in low areas such as pits, presenting a problem and requiring low-level ventilation for control.

Valves

Pressure Relief With gases, there is always the hazard of excessive pressures. Excessive pressures can result from exposure to sunlight and fires or from warming of cryogenic gas valves. To help alleviate this pressure so that the tank will not fail, pressure-relief valves are installed on compressed and liquefied gas tanks. These devices can be activated by pressure, heat, gas flow, or a combination of these. Containers of highly toxic gases do not have pressure-relief devices, since a prematurely operating or leaking device would present a greater hazard.

Pressure-actuated devices use the principle of rupture discs, spring-controlled relief valves, or hydraulic valves. As the pressure quickly rises, these devices actuate so that the container can handle the pressure. Temperature-actuated devices use fusible metal plugs or bimetal springs. These devices operate well at high temperatures or in fires. They are not effective for overfilling pressure increases (8).

Excess flow   Excess flow-actuated valves are designed to close when equipment failure releases a large volume of gas. The valves operate in a narrow pressure drop range when the leaking gas exceeds the predetermined flow rate of the valves. The best use comes at bulk storage terminals. Here trucks are mostly involved—they pull away from the terminal before disconnecting the hose and break the hose connection. If the flow is sufficient, the valve will operate to close the leak.

Emergency Shutoff   The emergency shutoff valve is normally located where the swivel-type or hose piping is connected to the fixed piping of the system. Temperature-sensing devices at the valve automatically shut the valve in the event of a fire. The valve is located on the fixed piping so that if a pull occurs on the swivel or hose part of the piping system, the valve will remain in place. The valve can be operated manually at the installed location or from a remote location.

Sources of Ignition

Ignition sources should be eliminated. Cutting and welding operations should be prohibited where flammable gases are being stored or produced. These operations provide an immediate ignition source from the open flames, sparks, or droplets of hot molten metal. Some other sources are friction sparks from metal-to-metal or metal-to-hard material contact, flames, arcs, hot surfaces, controls, appliances, smoking materials, and static electricity (1, 8).

The major container hazard for liquefied gases is the boiling liquid expanding vapor explosion (BLEVE). This is a type of pressure-release explosion where the container is heated by an outside source such as fire impingement. Even though container weakening is more often caused by fire, it can also occur when the container is punctured, bent, or damaged for other reasons. The liquid vaporizes with a large liquid-to-vapor expansion. This, combined with the weakening of the tank, causes tank failure. Failure occurs before the pressure-relief valve can relieve the pressure. The fireball from a BLEVE can be quite spectacular, due to the rapid mixing of vapor and air and the subsequent ignition by the fire that caused the BLEVE. The atomized droplets of liquid burn as they are propelled through the air. The magnitude of the BLEVE depends upon the weight of the container pieces and the amount of liquid that vaporizes. Insulated tanks designed for fire exposure conditions can help in protecting against BLEVEs. Water spray from hose streams or a fixed system entrained on the portion of the tank that is not in contact with the internal liquid will provide protection (1).

Safe Handling of Compressed Gases in Containers

The proper handling of compressed gas cylinders is essential to the safety of workers. The handling of cylinders is not hazardous in itself; however, failure to observe certain precautions can result in a variety of dangerous situations. The following practices should be observed when working with compressed gas:

Avoid accumulating more spare cylinders than you really need.

All cylinders should be properly secured in a vertical position to an immovable object (e.g., bench or wall).

Never drop cylinders or permit them to strike each other violently.

Valve protection caps should always be screwed into position on top of tanks in storage.

Oxygen cylinders should not be stored or used near flammable gases and solvents, and flammable gas cylinders should not be stored or used near sources of ignition.

Cylinders of gases should not be kept beyond the maximum recommended retention time specified by the supplier. A six-month retention period is recommended for (1) acid and alkaline gases that may corrode cylinders or cylinder valves, (2) gases subject to explosive decomposition, and (3) gases subject to polymerization.

Cylinders containing flammable gases should be electrically grounded before attaching equipment or opening valves.

Never open a gas cylinder valve that has no pressure regulator attached.

Pressure-regulator fittings are standardized to fit only cylinders containing gases with which they are compatible.

Regulators used for oxygen should never come in contact with oil, grease, other foreign matter, or an explosion may result.

Never tamper with safety devices on cylinders or valves.

Refer also to "Safe Handling of Compressed Gases in Containers," CGA Pamphlet P-1 (9), which recommends safe practices for users of gas products that are applicable to most industrial gases. Also refer to the *Handbook of Compressed Gases* (10) and the *Matheson Gas Data Book* (11).

It is very important that those who are involved with cylinder-filling operations know filling densities. *Filling density* is the percent ratio of the gas weight in a container to the weight of water that the container will hold at 60°F (15.5°C). Filling density is expressed as a percentage of the total water capacity, which is the amount of water, in either pounds or gallons at 60°F (15.5°C), required to fill a container. For LPG, this density can range from 48% to 57%, depending on the specific gravity of the LPG, the size of the container, and the container location (see Table 6.2). Maximum filling densities are specified in the U.S. Department of Transportation (DOT) regulations on hazardous materials (12) and in 29 CFR 1910.110 (5) for liquefied petroleum gas. These densities are necessary to ensure a vapor space above the liquid to allow for sufficient expansion if the temperature increases.

Specific Gases

Certain gases are encountered frequently in industry, and the specific characteristics and problems associated with these substances—principally liquefied petroleum gases, natural gas, acetylene, oxygen, and anhydrous ammonia—merit a more detailed discussion.

Liquefied Petroleum Gases

Liquefied petroleum gas is a volatile substance that figures in a number of fires, explosions, and injuries in the United States each year. As discussed earlier, LPGs are liquids when under pressure but vaporize upon the release of pressure; consequently they are stored and transported as liquids but used in their gaseous

Table 6.2 Maximum Permitted Filling Density[a]

| | Aboveground containers | | |
Specific gravity at 60°F (15.6°C)	0 to 1200 U.S. gals (1000 imp. gal, 4550 liters) total water capacity	Over 1200 U.S. gals (1000 imp. gal, 4550 liters) total water capacity	Underground containers all capacities
	Percent	Percent	Percent
0.496–0.503	41	44	45
0.504–0.510	42	45	46
0.511–0.519	43	46	47
0.520–0.527	44	47	48
0.528–0.536	45	48	49
0.537–0.544	46	49	50
0.545–0.552	47	50	51
0.553–0.560	48	51	52
0.561–0.568	49	52	53
0.569–0.576	50	53	54
0.577–0.584	51	54	55
0.585–0.592	52	55	56
0.593–0.600	53	56	57

[a]Courtesy of Occupational Safety and Health Administration.

states. These gases include propane, propylene, butanes, and butylenes. Some interesting properties of these gases are listed in Table 6.3.

Except when used as a raw material in a chemical reaction, LPGs are provided with an odorant as an aid to indicate the presence or leakage of an LPG. While these odorants, usually mercaptans, are flammable in themselves, they have no effect upon the properties other than to provide a distinctive odor to an otherwise odorless material. The odorant is excluded in LPGs used in certain manufacturing processes because of its possible adverse effect on the chemical reactions involved.

LPG fires have caused numerous deaths and, as a consequence, knowledgeable fire departments will approach this type of fire with extreme caution and may decide, justifiably so, not to approach within hose-stream reach under conditions where LPG containers are subject to fire exposure. Storage containers will rupture from heating and softening of the unwet crown, and while modern welded tanks may not rocket, this rupture results in the release of the liquid contents with flashing of the material into the vapor form, immediately creating an extremely high energy release and resultant fire and explosion. Under such condition, the operation of the relief valves will probably not be sufficient to prevent rupturing, although under light-exposure conditions, they may be expected to operate satisfactorily. It should be noted that discharges from tank relief valves have been known to ignite and form a blow torch of considerable dimension, exposing important facilities (see the earlier discussion on BLEVEs).

Industrial installations of LPGs will usually involve storage containers of up to 30,000-gal water capacity, although larger tanks are fabricated and in use. A typical LPG storage facility is shown in Figure 6.6. These installations should be de-

Table 6.3   Properties of Some Liquefied Petroleum Gases

Property	Propane	Butane	Isobutylene	Propylene (Propene)
Vapor pressure	124 psi at 72°F (22°C)	29.4 psi at 66°F (18.8°C)		147 psi at 68°F (19.8°C)
LEL	2.3	1.9	1.8	2.0
UEL	9.5	8.5	8.8	11.1
Vapor density	1.56	2.05	1.94	1.5
Boiling temperature	−43.6°F (−42°C)	31°F (−0.5°C)	19.6°F (−6.9°C)	53.9°F (−47.7°C)

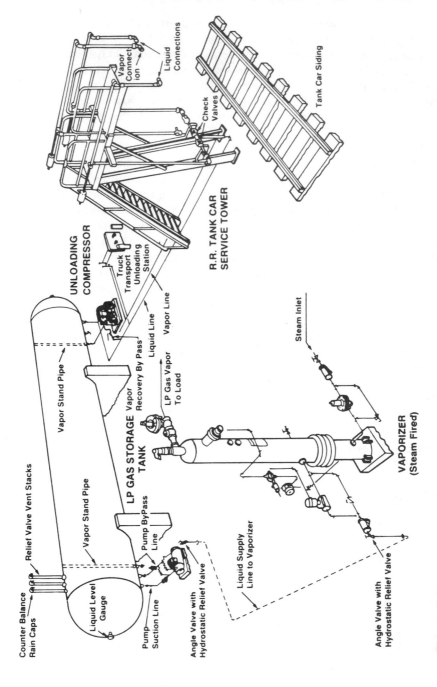

Figure 6.6 An example of an LPG storage facility. (Reprinted with permission from the Factory Mutual System.)

signed in accordance with the requirements of 29 CFR 1910.110, *Storage and Handling of Liquefied Petroleum Gases* (5), the *Standard for the Storage and Handling of Liquefied Petroleum Gases*, NFPA 58 (4), and the Unfired Pressure Vessel Section of the *Boiler and Pressure Vessel Code* sponsored by the American Society of Mechanical Engineers (13). Containers used in so-called bottled-gas systems (compressed gas cylinders) are constructed in accordance with DOT specifications (12).

Codes require that containers be marked in accordance with the following:

1. The code under which the container is constructed
2. Whether the container is designed for underground or above ground installation, or both
3. The name and address of the supplier of the container or the trade name of the container
4. Water capacity of the container in pounds or gallons
5. The pressure in psig for which the container is designed
6. Wording: "This container shall not contain a product having a vapor pressure in excess of _____ psig at 100°F"
7. With a tare weight in pounds or other identified unit weight for containers with a water capacity of 300 lb or less
8. With marking indicating the maximum level to which the container may be filled with liquid at temperatures between 20 and 130°F (–7 and 54°C), except on containers provided with fixed maximum level indicators or that are filled by weighing
9. The outside surface area in square feet

All container appurtenances (shutoff valves, check valves, excess-flow valves) should be fabricated of material suitable for LPG service and resistant to the action of LPG under service conditions. They should also have a rated working pressure of 250 psig minimum and should not be constructed of cast iron. Filling, withdrawal, and equalizing connections should be equipped with appurtenances depending upon the type and capacity of container and the service for which they are used.

Excess-flow valves currently in use are designed to be accessible from outside the tank and are provided with a shear groove to prevent damage to the valve from an external blow. Old-style types are installed inside the tank in such a manner as to prevent mechanical injury to the excess-flow valve in the event of damage to exterior piping.

These valves are designed to close at from 150% to 200% of a normal flow and must be installed in the correct position to enable them to close with the direction of flow. While not required by the code, these valves should be tested at 5-year frequencies for proper closing action by suppliers or those completely knowledgeable in the field of LPG equipment. In general, excess-flow valves should be provided at every pipe connection to the containers, with the exception of relief valve and filler connections.

The National Institute for Occupational Safety and Health (NIOSH), in a research study, determined that this type of valve influences the course of LPG-related fire incidents at liquid-transfer facilities (14). NIOSH used a logic diagram to analyze the case histories of LPG fires involving liquid-transfer operations. Typically, a liquid-transfer system consists of a storage tank or tanks, a pump, a dispenser (transfer station), a piping system, and various container appurtenances such as excess-flow valves, back-flow check valves, and safety relief devices (Fig. 6.7). This typical layout served as the basis for the development of the logic tree diagram, which is shown in Figure 6.8. The diagram contains the three components of the fire triangle: heat (ignition source), oxygen supply, and fuel. The fuel component, "propane supply from main tank," is separated into six output events or failure points that correspond to the basic layout in Figure 6.7. The potential failure source that could occur at each of the six areas include employee operating errors, inadequate equipment design or specifications, valve failures, and other equipment failures. Figure 6.8 contains only the initial portion of the complete logic tree. This logic tree was then applied to the actual LPG fire accident data to see what a structured analysis could reveal.

From this study, NIOSH concluded the following:

1. Excess-flow valves are good for controlling high-volume leaks; but for low-volume leaks, emergency shutoff valves in addition to excess-flow valves are essential.
2. Reaffirmed the NFPA requirement that recommends the installation of emergency shutoff valves protected by a concrete bulkhead or equivalent anchorage.
3. Frequent inspection and maintenance of control valves and of all equipment are crucial.
4. Future efforts should be directed toward educating the distributors, since they have the ultimate responsibility for the safe distribution and usage of LPG. Better methods of user education are vital.

In addition to the results of this research, the logic tree diagram method should prove to be a valuable tool to use in the fire loss control program for investigating other types of operations.

Back-pressure check valves should be provided on fill connections and may be of the spring-loaded or weight-loaded type with in-line or swing action designed to close when flow is either stopped or reversed. Internal valves that are installed so that damage to the exterior will not prevent operation of the valves, that are

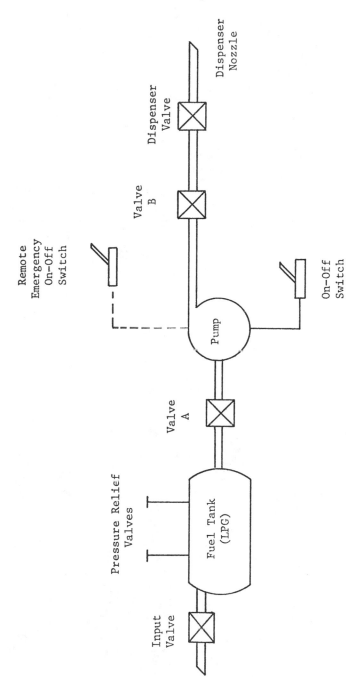

Figure 6.7 Diagram of a basic system for dispensing liquified petroleum gas. (Courtesy of National Institute for Occupational Safety and Health.)

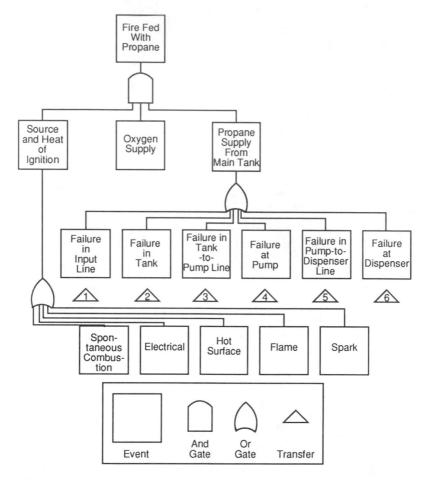

Figure 6.8 Logic tree diagram (Courtesy of National Institute for Occupational Safety and Health.)

either manually or automatically operated, and that are designed to remain closed except during operating periods are considered positive shutoff valves.

Emergency shutoff valves should be installed and should incorporate all of the following: (1) automatic shutoff through thermal actuation, (2) manual shutoff from remote location, and (3) manual shutoff at installed location.

Safety relief valves are obviously vital for all container installations used for storage of LPG. For large, ASME-designed containers, these valves are spring-loaded, with small DOT-type containers (bottle-gas type) equipped with either safety relief valves or fusible-plug safety relief devices. Spring-loaded valves for containers are generally required to be set to start to discharge at a minimum of 88% and a maximum of 100% of the design pressure of the container. A plus tolerance not exceeding 10% of the set pressure marked on the valve is permitted. Do not install shutoff valves between the safety relief valve and the container unless a three-way valve is installed beneath two safety relief valves, permitting one valve to be fully open at all times or, where separate safety relief valves are provided, interlocked to ensure that one valve is open at all times. These situations, which are exceptions, are frequently required in industries where safety relief valves are subject to clogging and require frequent servicing or where processing cannot be interrupted for testing.

Each safety relief valve should be plainly and permanently marked with the following:

1. The container type on which the valve is designed to be installed
2. The pressure in psig at which the valve is set to discharge
3. The actual discharge rate in ft^3/min of air at 60°F (15.6°C) and 14.7 psia
4. The manufacturer's name and catalog number

U.S. DOT specification cylinders should be provided with fusible plugs, spring-loaded safety relief valves, or a combination of both for relief of excessive pressures. Plugs may be used only on aboveground stationary containers having a maximum volume of 1200 gal water capacity.

For containers filled by volume (this includes all but the smaller DOT cylinders, which are filled by weighing), liquid-level gauging devices are required with fixed-level gauges or variable gauges of the slip-tube, rotary-tube, or flow types (or combinations of gauges). Those gauging devices that require bleeding of the product to the atmosphere, such as fixed-liquid-level, rotary-tube, and slip-tube gauges, must be designed so that the bleed valve maximum opening to the atmosphere is not larger than a No. 54 drill size unless equipped with an excess-flow check valve.

Vaporizers are frequently needed, particularly for LPGs having lower vapor pressures, such as butanes. Indirect-heated types using steam and hot water are desirable, although direct-gas-fired types are permissible and commonly encountered, principally for economic reasons. These vaporizers should be located outside or in noncombustible vaporizer houses. Indirect types should have the heating unit separated by a vapor-tight, noncombustible partition.

Small DOT specification cylinders containing LPG are normally from 5- to 125-gal water capacity and are not, as a matter of general practice, refilled on the user's property. Cylinder systems should preferably be located outside buildings, a minimum of 5 ft from windows or any other openings. The use of portable containers is permitted on roofs or inside buildings or structures if connected for use where outside locations are impractical, provided the containers do not exceed 245-lb water capacity each. Cylinders located inside buildings should only be in well-ventilated areas and, if possible, should be in locations provided with sprinkler protection. Such protection has been shown by actual experience to control LPG fires by keeping containers cool pending operation of the fusible plugs or safety relief valve.

Regulators or pressure-reducing valves are used to control distribution or utilization pressure and should be located as close to the container as practical. First-stage regulating equipment should be located outside buildings. For regulating equipment installed outside buildings, the discharge from safety relief valves should be located not less than 3 ft horizontally away from any building opening below the level of such discharge, nor beneath any building unless the space is well ventilated and not enclosed for more than 50% of its perimeter. On regulators installed inside buildings, the discharge from the safety relief device and the spaces above the regulator and relief valve diaphragm should be ventilated to the outside, again not less than 3 ft horizontally away from any building opening below the level of such discharge. It is generally desirable to vent the diaphragm chamber to the outdoors in the event of rupture, unless the interior location is large and well ventilated. It is permissible to pipe LPG liquid or vapor at all normal operating pressures outside buildings. However, pressures in buildings should not exceed 20 psig. Exceptions to this are buildings and separate areas of buildings used to house equipment for vaporization, pressure reduction, gas mixing, gas manufacturing or distribution, or industrial processes; research and experimental laboratories or equipment or processes having a similar hazard; buildings or structures under construction or undergoing major renovation; or buildings or structures in which liquid feed systems are used, provided liquid piping is of heavy-wall, seamless brass or copper tubing not exceeding 3/32-in. internal diameter with a wall thickness of not less than 3/64 in.

Whenever possible, piping should be located aboveground whether in or exterior to buildings. Leaks in underground piping are difficult to locate. Piping buried under buildings or inaccessible crawl spaces should be avoided. Aboveground piping also must be protected from physical damage. Where buried pipe cannot be avoided, it should be protected against corrosion.

In addition to safely locating pipe, further consideration is needed in selecting the proper weight and material to be used. Piping used for pressures in excess of 125 psig should be Schedule 80, and piping used for pressures under 125 psig should be Schedule 40 or heavier. Flexible connectors should be used where necessary; however, the use of nonmetallic hose for permanently interconnecting containers should be avoided. Type K or L copper tubing may be used in LPG service, provided the soldering or brazing filler material has a melting point exceeding 1000°F (538°C), with flared tubing joints also acceptable. Use approved hose designed for a minimum bursting pressure of 1750 psig. This hose is marked "LPG" at intervals not greater than 10 ft for ready identification.

All new gas piping or piping that has undergone repair should be tested at 1-1/2 times the normal working pressure, but at not less than 3 psig, with an inert gas or air used as a testing medium.

Sources of ignition must be eliminated in all areas where there would normally be a release of LPG or where an upset condition might produce a release of the gas. Open flames and spark-producing equipment such as ordinary portable electrical tools and extension lights should not be permitted. Electrical equipment and wiring should be installed in accordance with the provisions of 29 CFR 1910, Subpart S (5) and the *National Electrical Code,* NFPA 70 (4) for ordinary locations, except that fixed electrical equipment in classified (hazardous) areas should comply with the requirements of 29 CFR 1910.110 (5) and NFPA 58 (4).

It is to be noted that static electricity is not a problem in LPG installations, because of the closed nature of the systems and the non-static-producing characteristics of gas flow through piping. However, it is suggested that at liquid transfer points, static grounding and bonding be provided in a manner similar to that for flammable liquid transfer systems.

Dry chemical is a very effective extinguishing medium for LPG fires and for gas fires in general. As a minimum, a portable dry-chemical extinguisher of adequate capacity should be provided at each LPG installation. An LPG fire should not be extinguished unless it is possible to shut off the flow of gas. Extinguishing and permitting the continued flow of the gas with the possibility of vapors reaching a source of ignition will generally result in a much greater area of involvement and loss. Thus, controlled burning can be desirable in many cases.

In multiple-container installations, adequate water in the form of hose streams from a nearby hydrant or hydrants should be available to play upon the unwet crown (portion of the tank not in contact with internal liquid) of the storage containers to prevent rupture and involvement of all containers. Overpressurization and rupture of containers, as attested to by loss records, can and do result in disasters of catastrophic proportions.

As stated previously, fire fighters are understandably cautious in approaching a fire at an LPG facility because of the extensive loss of life sustained in combating such incidents. The need for the use of hose streams will be greatly reduced, but not eliminated, by

1. Insulating the tanks
2. Mounding or burying the tanks
3. Providing automatic fixed water-spray systems

Insulation of the tanks, using a noncombustible insulation to limit the temperature of the metal in the unwet crown to 800°F (427°C) after 2 hr of exposure to an average temperature of 1800°F (982°C), is recommended. The insulation will require little or no maintenance, as compared to the maintenance required for the water-spray system and single failure point possibilities of fixed automatic equipment. Burying and mounding of the tank will require corrosion protection and present problems in subsequent tank inspections.

The location of LPG containers with reference to important buildings, electrical substations, boiler houses, exterior process units, vital cooling towers, etc., should provide for maximum separation to avoid exposure damage to important facilities. In addition to adequate separation, it is also highly desirable to orient the axis of LPG containers so that the tanks are parallel to and not pointed at important facilities or, as indicated above, parallel to ancillary structures or equipment vital to business continuity.

OSHA 29 CFR 1910.110 (5) requires minimum clearances of LPG containers from important buildings or property lines. Table 6.4 lists these minimum distances for containers of various sizes and also for the aggregate capacity of grouped containers.

Current energy problems, and specifically the shortage of natural gas, are compelling many industries to add LPG storage installations. Frequently, the facilities at which LPG is urgently needed are congested to the degree that the minimum container distances cannot be met. Where this problem exists, either a free-standing fire wall should be constructed or one of the methods mentioned previously (insulation of tanks, burying or mounding, or fixed, automatic water-spray systems) should be employed (with approval of the authority having jurisdiction). Fire walls, to be acceptable, should extend slightly above the height of the container, about 1 ft, but where tanks are exposing important buildings, particularly if multistory, this height may not be adequate. Freestanding walls by necessity will be expensive and preferably should be reinforced concrete. Fire walls or the other means of protection should also be used where minimum distances cannot be provided between groups of tanks.

With the grouping of tanks, a minimum of 25 ft between these installations is required. For capacities greater than those listed in Table 6.4, the authorities having jurisdiction may require fire walls or other means of special protection previously mentioned.

Natural Gas

Because of the problems presented by LPG and specifically its vapor being heavier than air, our discussions to this point have been mainly concerned with this material rather than with natural gas, which is used extensively throughout the country. We will not attempt to devote a great deal of time to the topic of natural gas other than to point out that problems generally occur from the escape of the gas from piping systems or from its use as a fuel, which will be covered by a separate area of this chapter. For a detailed treatment of this subject, refer to the *Na-*

tional Fuel Gas Code, NFPA 54 (4). Seldom are high pressures used for natural gas. Natural gas requires the use of standard weight, Schedule 40 steel or wrought iron pipe for gas pressures up to 125 psig. Copper and brass pipe and tubing are also acceptable, but copper pipe or tubing is desirable for buried pipe with service pressures up to 100 psi because of its ease of installation, ductility, and resistance to corrosion. Cast iron should not be used (as pointed out before) because of its relative ease of fracture, and copper should not be considered for service where gas contains more than an average of 0.3 grain of hydrogen sulfide per 100 ft^3 of gas.

It is recognized that much natural gas piping is buried underground, which has resulted in many catastrophes due to leaks caused by corrosion or rupture owing to physical damage or poor installation procedures. Leaking has has been known to travel considerable distances before finding a source of ignition, generally in

Table 6.4 Distances of Containers to Important Buildings[a]

Water capacity per container	Containers Underground	Containers Aboveground	Between aboveground containers
Less than 125 gals[b]	10 ft	None	None
125 to 250 gals	10 ft	10 ft	None
251 to 500 gals	10 ft	10 ft	3 ft
501 to 2,000 gals	25 ft[c]	25 ft[c]	3 ft
2,001 to 30,000 gals	50 ft	50 ft	5 ft
30,001 to 70,000 gals	50 ft	75 ft[d]	
70,001 to 90,000 gals	50 ft	100 ft[d]	

[a]Courtesy of Occupational Safety and Health Administration.
[b]If the aggregate water capacity of a multicontainer installation at a consumer site is 501 gallons or greater, the minimum distance shall comply with the appropriate portion of this table, applying the aggregate capacity rather than the capacity per container. If more than one installation is made, each installation shall be separated from another installation by at least 25 feet. Do not apply the MINIMUM DISTANCES BETWEEN ABOVEGROUND CONTAINERS to such installations.
[c]The above distance requirements may be reduced to not less than 10 feet for a single container of 1200 gallons water capacity or less, providing such a container is at least 25 feet from any other LPG container of more than 125 gallons water capacity.
[d]One-quarter of sum of diameters of adjacent containers.

basement areas. Where possible, as with LPG in a plant facility, it is desirable to locate the piping aboveground, properly supported and protected from physical damage. Where gas must be run in earth or other material that may provide corrosion problems, piping should be protected against corrosion. Particularly undesirable are locations having earth with a high cinder content. Piping underground beneath buildings should be avoided. However, if piping must be laid beneath the building, the piping should be encased in a conduit sealed to prevent the possible escape of any gas leakage. Where piping enters the building through the foundation wall, it should be encased in protective piping, with the annular space between the gas piping and the sleeve sealed at the foundation or basement wall to prevent entry of gas or water. The installation of gas piping should be avoided, wherever possible, in inaccessible locations, such as crawl spaces.

All piping should be pressure tested prior to initial operation to ensure tightness. The presence of leaks should be determined by means of a soap-bubble test or other foaming agent. A reduction of test pressures as indicated by pressure gauges is an indication of the presence of a leak.

Gas meters should be located in ventilated spaces that are readily accessible and in locations not subject to physical damage or corrosion. Gas meters should be located at least 3 ft from sources of ignition or where they will not be subjected to excessive temperatures or extreme changes in temperature. Frequently, large meters are located in boiler rooms, in main areas, etc., which, while acceptable, cannot be considered good practice.

Overpressure protection devices should be provided when failure of the pressure regulator could provide downstream pressures that might result in hazardous conditions or damage to equipment. These overprotection devices will take the form of pressure-relieving devices, automatic shutoffs, or regulators. Regulators are required when the gas supply pressure is higher than that at which the gas equipment is designed to operate or varies beyond the pressure limits designed into the equipment. Pressure regulators are equipped with diaphragms that can rupture. Consequently, they should be vented to a safe location outside the building. Under no circumstances should they be vented to the gas equipment flue or exhaust system.

Liquefied Natural Gas

Liquefied natural gas (LNG) is classified a cryogenic gas and is natural gas composed mainly of methane with small amounts of ethane, propane, and butane. This gas is used both as a source of natural gas to augment pipeline supplies during peak demands and as a basic supply of natural gas. LNG is shipped in insulated marine vessels and tanker trucks designed to DOT specifications. LNG is in a liquefied state at $-260°F$ ($-162°C$) and, upon release to the atmosphere, will vaporize at ambient temperature with an expansion rate of 600 times the volume of the liquid vaporized. At temperatures below approximately $-170°F$ ($-112°C$),

LNG is heavier than ambient air. At 60°F (16°C), however, as the temperature rises, LNG then becomes lighter than air. For further information, consult the *Standard for the Production, Storage, and Handling of Liquefied Natural Gas (LNG)*, NFPA 59A (4).

Acetylene

Acetylene is widely used as a raw material in the chemical industry in acetylenic chemistry and also in welding and cutting. Chemically pure acetylene is colorless and odorless, with an ignition temperature of 571°F (299°C), a wide flammable range of 2.5% to 81%, and a vapor density of 0.9. Decomposition of acetylene occurs at 15 psi without the presence of air, evolving hydrogen and carbon with explosive force if ignited. As a consequence, acetylene generators are operated at below this figure. The sensitivity of acetylene to explosion is resolved in cylinders by dissolving compressed acetylene in acetone with a special porous filler.

Acetylene is generally produced by the action of calcium carbide and water or by re-forming hydrocarbons such as methane, propane, and naphtha. Acetylene generators are occasionally found in industry and may be of carbide-to-water or water-to-carbide types. Generators in common use are of the carbide-to-water type, principally for safety reasons. This type will more readily dissipate the heat produced by the reaction by controlling the rate at which the carbide is allowed to drop into the water. Other types of generators are the water-recession type, where water is brought in contact with the carbide by change of water level with the water level held back by gas pressure, and the water-to-carbide type, where water in limited quantity is brought in contact with the main mass of carbide.

Later in this chapter is a discussion of oxygen-fuel gas systems, including acetylene. However, a few comments on the handling of acetylene cylinders are in order. Store, secure, and use acetylene cylinders in an upright position. Close valves on empty cylinders while in storage and during shipment. Store cylinders at least 20 ft from highly combustible materials inside buildings or, for that matter, outside buildings. Use caps for valve protection when cylinders are not in service or connected for service.

Oxygen

Although nonflammable, oxygen cannot be omitted from this discussion because it is a supporter of combustion to such an extent that certain materials such as oils and other combustible lubricants may ignite explosively at ordinary temperatures in the presence of pure oxygen. Many materials thought to be only slightly combustible will also burn intensively in oxygen, and the flammable characteristics of other materials, such as flammable liquids, will be greater in oxygen than in air, with wider flammability limits, etc.

Oxygen is generally obtained by the liquefaction of air (which also produces nitrogen). It is handled as a gas or in a liquid form, with liquid oxygen storage becoming more common because the volume expansion ratio from liquid at 1 atm to gas at atmospheric pressure is approximately 1:862. This has practically eliminated the use of high-pressure oxygen systems at facilities using large quantities of oxygen for welding and cutting or in medical institutions.

Oxygen-Fuel Gas Systems

The use of oxygen and various fuel gases such as acetylene, liquefied petroleum gas, natural gas, hydrogen, and methylacetylene-propadiene (MAPP) is extensive largely in welding and cutting operations and to a lesser degree in heat-treating and other processing. Previous sections have covered the hazards of some of these gases. Combining these materials in industry requires preventive action to avoid the intermixing of a fuel gas with oxygen and the propagation of an oxygen-fuel gas flame back to the fuel gas source. Oxygen, while itself nonflammable, is a supporter of combustion, and mixing it with flammable gases creates unusually high temperatures and energy release during combustion. Flammable gases used with oxygen will be referred to in this section as *Fuel gases*. Reference should be made to 29 CFR 1910, subpart Q, *Welding, Cutting, and Brazing* (5), and the *Standard for the Design and Installation of Oxygen-Fuel Gas System for Welding, Cutting, and Allied Processes*, NFPA 51 (4) for more requirements and information. Welding and cutting procedures, including the use of hot-work permits are covered in Chapter 3, Elements of a Fire Loss Control Program and Chapter 7, Common Process Hazards.

One of the most frequent areas cited by OSHA compliance officers is the storage of oxygen and fuel gas cylinders. These cylinders, while they may be permitted inside buildings, should be stored a minimum of 20 ft from combustible materials; in addition, they should not be exposed to excessive rise in temperature, physical damage, or tampering. Cylinders should also have their valves closed when in storage and during shipment and caps should be in place except when cylinders are in service or connected ready for service. Obviously, cylinders in storage or service should be adequately secured to prevent toppling. Fuel gas cylinders stored in the building, with the exception of those in use or attached ready for use, should be limited to a total gas capacity of 2500 ft³ of acetylene or nonliquefied gas or to a total water capacity of 735 lb of liquefied petroleum gas or methylacetylene–propadiene, stabilized (MPS).

Fuel gas storage within a building exceeding the limitation should be stored in a separate room of noncombustible (or limited combustible) construction, with a minimum fire resistance rating of 1 hr, with walls or partitions continuous from floor to ceiling and securely anchored, and with a minimum of one wall of the room an exterior wall. Openings from the inside room should be protected by approved minimum 1-hr rated fire doors of the swinging, self-closing type. Windows and partitions of wired glass should be in approved metal frames with fixed sash. Explosion venting, with a minimum ratio of not less than 1 ft^2/50 ft^3 of room volume, designed to relieve a maximum pressure of 25 lb/ft^2, should be provided.

Inside storage rooms should be well ventilated with vents located at floor and ceiling. Electrical equipment of Class I, Division 2 should be provided in these areas.

To prevent an extremely intense and high-energy-release fire involving oxygen and fuel gas, oxygen cylinders in storage should be separated, not only from the fuel gas cylinders, but also from combustible materials where a fire in an oxygen atmosphere would also be intense. Consequently, no oxygen cylinders are permitted in acetylene generator enclosures, and cylinders should be separated from generators or calcium carbide storage by a noncombustible partition having a minimum fire resistance rating of 1 hr with gas-tight partitions without openings. Oxygen and fuel gas cylinders should be separated from each other by at least 20 ft or by a noncombustible barrier, a minimum of 5 ft high, having a fire resistance rating of at least 0.5 hr.

Limitations should also be imposed upon the quantity of oxygen and fuel gas manifolded for use. For inside manifolds, fuel gas cylinders should be limited to a total gas capacity of 3000 ft^3 of acetylene or nonliquefied gas or a total water capacity of 735 lb for LPG or MPS. It is permissible to have more than one manifold in the same room, provided these manifolds are separated by a minimum of 50 ft. Fuel gas cylinders connected to a manifold and having capacities in excess of those listed above should be located either outside the building or in a separate building or room, with enclosure requirements the same as those listed above for storage of fuel gas cylinders.

As with separation of oxygen and fuel gas cylinders in storage, oxygen manifolds should also be located a minimum of 20 ft from fuel gas cylinders or be separated in a manner similar to the storage separation. The limitation of oxygen capacity for one manifold is a maximum total gas capacity of 6000 ft^3, with more than one such manifold permitted to be located in the same room, provided the manifolds again have a minimum 50-ft separation. For oxygen manifolds in excess of 6000 ft^3, locations outside or in a separate room with the same requirements as those mentioned above for indoor gas storage should be used. These requirements are for high-pressure oxygen manifolds with service pressure above 250 psig. For bulk oxygen supplies, refer to the *Standard for Bulk Oxygen Systems at Consumer Sites*, NFPA 50 (4), and 29 CFR 1910.104 Oxygen (5).

For low-pressure manifolds having service pressure not exceeding 250 psig, the requirements for manifold locations are the same as those for high-pressure manifolds. A sign on the installation should indicate that it is a low-pressure manifold with caution not to connect high-pressure cylinders to a manifold having a maximum pressure of 250 psig.

Because of the possibility of acetylene decomposition pressures in excess of 15 psig, this gas should not be piped at pressures in excess of this amount and, in addition, the piping should be steel. Reduction of pressure to this limit is accomplished by the use of pressure regulators at cylinders or manifolds. Unalloyed copper is prohibited except when contained in listed equipment, because acetylene in contact with copper will form explosive substances. Oxygen piping should be at least Schedule 40 and tubing of Types K or L. Materials used in piping oxygen may be of steel, brass, or copper, and tubing of seamless copper, brass, or stainless steel may also be used. However, oxygen at pressures in excess of 700 psig should be handled in stainless steel or copper alloys.

Oxygen piping considerations include the installation of readily accessible gas valves at buildings and valves in the discharge from manifolds or other sources of supply. Where extensive fuel gas piping systems extend through plant areas, consideration should be given to the installation of additional valves for sectional control for ready emergency shutoff in the event of leaks or damage. It is advisable to identify the valves by placards to facilitate prompt response during emergencies.

Protective equipment designed, installed, and used for the service for which it is approved is required for pressure relief, to prevent a back flow of oxygen into the fuel gas supply system and also to prevent passage of a flame back to the fuel gas supply. These devices take the form of relief valves, check valves, and hydraulic flame arresters with check valve and pressure relief. Back-flow prevention should be installed downstream of the fuel gas and oxygen station outlet valve, but other protective equipment may be installed at any point downstream of the fuel gas supply.

A check valve, pressure regulator, hydraulic arrester or seal, or combination of these devices should be provided in each station outlet, with station outlets also required to be provided with a detachable outlet seal cap secured in place except when a hose regulator or piping is attached. Experience indicates that a hydraulic arrester will provide the greatest degree of safety, but liquid levels should be checked at frequent intervals.

The use of approved pressure-reducing regulators is required for fuel gas, attached to either the cylinder valve or the manifold and also at the oxygen cylinder to ensure a uniform gas supply to the torch at a proper pressure. Hose, for both oxygen and fuel gas service, is generally colored red for fuel gas and green for oxygen and should be approved for such usage, with parallel lengths of oxygen and fuel gas hose, when taped together for convenience, exposing a minimum of 4 in. of the hose out of each 8 in. Metal-clad or armored hose should not be used.

To prevent the connection of oxygen to fuel gas and vice versa, right-hand threads are used for oxygen regulators and left-hand threads for fuel gas regulators. It is important to use only regulators approved for the gas for which they are intended.

This section has attempted to highlight the requirements for safe handling of oxygen-fuel gas systems. Continuing onstream safe procedures used in cutting and welding and other operations involving the use of oxygen-fuel gas, such as testing of hose, obtaining a welding and cutting permits, etc., are also a necessity for fire safety when handling these materials. The criteria of primary importance in handling oxygen is adequate separation of oxygen from flammable gases and other combustibles.

Anhydrous Ammonia

Before closing the section on gases, mention should be made briefly on anhydrous ammonia, which is used extensively throughout the industrial and agricultural world. Many think that because this material is shipped as a nonflammable gas, there is no hazard. However, under certain conditions, ammonia can be hazardous if it is in concentrations above the lower explosive limit of approximately 16%.

Ammonia is easily liquefied by pressure only and has the properties of a colorless gas with a boiling point of $-28°F$ ($-33.5°C$) and a vapor pressure of the liquid of 8.5 atm at $68°F$ ($20°C$) and is very soluble in water.

Ammonia can be explosive if combined with oils under certain conditions, but the main hazard is the fact that it is a flammable gas when found in high concentrations. It might be added at this point that it is shipped as a nonflammable gas because of U.S. DOT regulations that consider gases with a lower explosive or flammability limit of 12% or higher as nonflammable (12).

CHEMICALS

An effective fire loss control program when chemicals are involved requires in-depth knowledge of the hazardous properties of chemicals. Since there are over 60,000 known chemicals in industry, with new chemicals being introduced every year, it becomes a major problem for individuals to identify and understand the hazards of these chemicals. Therefore, it is useful to classify chemicals according to hazardous properties: (1) ability to oxidize other chemicals, (2) combustibility, (3) reactivity with air or water, (4) instability, (5) corrosiveness, (6) radioactivity, and (7) toxicity. Caution should be exercised, because some chemicals possess more than one of these properties and some chemicals exhibit hazardous properties not covered by this general classification (e.g., blasting mixtures are classified as oxidizable substances, but are also explosives). Reference should be made to chemical dictionaries, such as the *Condensed Chemical Dictionary* (15); material safety data sheets (MSDS) provided by chemical manufacturers; chemical references, such as the *Hazardous Chemicals Desk Reference* (16); various

NFPA publications, including *Hazardous Chemical Data*, NFPA 49, *Manual of Hazardous Chemical Reactions*, MFPA 491M, and *Fire Hazard Properties of Flammable Liquids, Gases, and Volatile Solids*, NFPA 325M (4); and other similar sources of information.

Oxidizers

Most oxidizing chemicals are not combustible, but they increase the intensity of burning of combustible materials. Oxidizing chemicals should be stored away from flammable liquids and combustible materials.

Nitrates

Inorganic nitrates are used in fertilizers, salt baths, and other operations. They tend to release oxygen in a fire and can react violently with organic materials. Some common nitrates are sodium nitrate, potassium nitrate, ammonium nitrate, and cellulose nitrate.

Since most nitrates can explode and some, such as ammonium nitrate, detonate, careful consideration should be given to handling these materials and controlling fires involving these materials. In most cases, water appears to be the best extinguishing agent, but extreme care should be taken in the application.

Inorganic Peroxides

Most inorganic peroxides can ignite or explode with water, and some, such as hydrogen peroxide in concentrated solutions (greater than 92%), can be detonated by shock. Other common inorganic peroxides include sodium peroxide, barium peroxide, and potassium peroxide. Dry chemical extinguishing agent, dry sand, or soda ash can be used on peroxide fires.

Other groups of oxidizers are nitric acid, nitrates, chlorates, chlorites, dichromates, hypochlorites, perchlorates, permanganates, and persulfates.

Combustible Chemicals

Basically, all organic chemicals are combustible. Examples of some of the combustible chemicals are carbon black, lamp black, nitroaniline, nitrochlorobenzene, sulfides, sulfur, and naphthalene. In particular, sulfur forms highly explosive and detonatable mixtures with chlorates, perchlorates, and gun powder, when mixed with potassium nitrate and charcoal.

Water-Reactive and Air-Reactive Chemicals

These chemicals pose a substantial fire loss control problem because they react with water or air to liberate significant quantities of heat, can self-ignite, and can ignite other combustible materials. These chemicals include alkalies, aluminum trialkyls, anhydrides, carbides, charcoal, coal, hydrides, oxides, phosphorus, sodium hydrosulfite, and alkali metals (lithium, potassium, and sodium). These chemicals require special fire extinguishing methods, depending on the chemical.

Unstable Chemicals

Unstable chemicals are characterized by spontaneous polymerization, decomposition, and self-reaction. Some of these chemicals are acetaldehyde, ethylene oxide, hydrogen cyanide, methyl acrylate, methyl methacrylate, nitromethane, organic peroxides, styrene, and vinyl chloride. Special storage and fire-fighting procedures are necessary, depending on the chemical.

Corrosive Chemicals

Corrosive chemicals cause damage to materials and have a destructive effect on living tissue. Some strong oxidizers and water- and air-reactives are also corrosive.

Inorganic Acids

Inorganic acids can react with other chemicals or combustible materials to cause fire or explosions. Common acids are hydrochloric, hydrofluoric, nitric, perchloric, and sulfuric. Water spray is recommended for fighting fires in acid storage areas.

Halogens

Of the halogens, fluorine gas can be particularly troublesome, since it can combine with most known elements to create heat and possibly fires and explosions. Other halogens are bromine, chlorine, and iodine.

Radioactive Substances

These chemicals or substances, in addition to their normal fire and explosion hazards, have the added hazard of radioactivity. Contamination can spread during a fire to the smoke and to the water runoff during fire-fighting operations. It is usually preferable to have automatic sprinklers control this type of fire rather than manual fire fighting. More information can be obtained from the *Recommended Fire Protection Practice for Facilities Handling Radioactive Materials*, NFPA 801 (4).

Toxic Chemicals

The problem with toxic chemicals in a fire is the threat to people exposed in the fire area. Automatic fire protection is advisable for storage areas of highly toxic chemicals.

One of the most important things to remember about chemicals is to separate chemicals from incompatible chemicals and materials: acids from bases, oxidizers from flammables, unstable chemicals from everything, and so forth. Plus, the fire loss control managers should know the properties of the chemicals stored and used in their facilities and make use of this knowledge in their fire loss control programs.

In addition to knowing the chemical properties of materials, a determination of the fire and explosion hazard involves the following:

An extensive knowledge of the nature of the raw materials, or so-called *feedstock*, how those materials are combined, and the chemical processing involved, including such operations as oxidation, hydrogenation, nitration, etc.

An extensive knowledge of the physical operations involved, known as *unit operations*, which include filtration, distillation, etc.

The ability of operating personnel to avoid errors that can lead to unsafe conditions and their ability to effect a safe shutdown when unsafe situations occur. This is a human-element function of training and experience.

Knowledge of the actual physical plant layout, including site, climatic factors, and protection features.

COMBUSTIBLE SOLIDS

In considering combustible solids, the problem of dust explosions must be given high priority because of their potential for causing extensive property damage and employee injury. While many solid materials evidence varying degrees of flammability, any combustible material—and some not considered combustible by the average individual, such as various metals when in a finely divided state—can be subject to explosions, creating extremely rapid rates of pressure rise and high pressures. Different materials will evidence different rates of pressure rise and maximum pressures developed. However, in general, good dust control procedures will cover good practices for virtually all materials in this category.

Although dust explosions are infrequent, when they do occur, heavy property loss and, of even more concern, employee fatalities and injuries may be encountered. In most cases, dust explosions can be prevented or the impact considerably lessened by good engineering design incorporating the following:

Elimination or control of ignition sources

Proper design of the structures and equipment to withstand or relieve the force of an explosion

Elimination of dust accumulations by providing dust control at points of emission

Avoidance of undesirable structural features such as ledges, flat and wide duct work at ceilings, etc., where dust can accumulate

Good housekeeping to remove any dust before significant accumulations become apparent in areas where adequate engineering features are not incorporated

Good housekeeping to remove any dust before significant accumulations become
    apparent in areas where adequate engineering features are not incorporated
Consideration of inerting or explosion suppression systems where, despite pro-
    viding both good engineering design and loss prevention control, explosions
    still persist

In considering the elimination of ignition sources, the obvious ignition source
of an open flame, such as encountered in welding and cutting, is sometimes overlooked. In dust areas, it is absolutely essential that good cutting and welding procedures be enforced. The use of approved and proper electrical equipment for
dust areas is also essential with the National Electrical Code, NFPA 70 (4), requiring Class II, Division 1 equipment where dust may normally be expected to
be suspended in air. In locations where free dust will not be involved except under
upset conditions, Division 2 equipment will suffice. Static electricity, unless removed by grounding, is a source of ignition of dust explosions, and good static
removal practices should be followed.

As discussed earlier, it is economically impractical to design a structure to
withstand the anticipated pressures developed by most combustible dusts. As a
consequence, explosion-relief venting is employed either by providing so-called
damage-limiting construction (which is basically construction of a light, noncombustible nature, such as an all-metal building), or by providing sections of
walls or roof that will relieve at relatively low pressures (in the area of 20 to 25
lb/ft^2), preventing maximum pressure buildups within the structure. Where it is
impractical to design actual dust-reducing equipment, such as dust collectors to
withstand the force of an explosion, explosion-relief venting to the atmosphere
should be provided.

The location of a dust-producing process should be considered with care. If it
is not possible to locate these operations in a detached building, they should be in
a one-story section, not exposing upper stories of a main area, and should be separated by explosion-resistant walls. Under no conditions should a hazardous dust-
producing process be located in lower stories of a multistory building or, for that
matter, in basements where it is virtually impossible to obtain an adequate explosion-relief-venting ratio.

In good design, flat, wide ducts should be avoided in dust areas where dust will
accumulate and may be rather difficult to remove, depending upon ceiling
heights. These two considerations may seem rather obvious, yet they are frequently overlooked by architects and engineers. As mentioned in an earlier chapter, the evolution of dust explosions frequently begins as relatively minor puffs
dislodging dust from overhead building members and resulting in successive explosions, each more damaging than its predecessor. This condition can exist in
and is fostered by building interiors that are not suited to dust operations and that
have ledges and other flat surfaces where dust accumulations may occur. These

locations may be rather inaccessible, thus discouraging the frequency of cleaning needed to keep dust levels to a minimum.

Enough emphasis cannot be given to providing an adequate level of house-keeping and to removing any accumulations of dust that may escape even from well-designed equipment. What may appear to be rather thin layers of dust on equipment (in the area of 1/8 to 1/4 in. thick) can be extremely hazardous and should not be permitted to accumulate. This may require a frequent cleaning schedule where capture of dust emissions at points of liberation by dust-collecting systems is poor.

While an extremely wide range of materials have been involved in numerous dust explosions over the years, some of the materials that have been more prominently involved are grains, starches, spices, wood flour or dust, plastics, and aluminum powders. The ignition and intensity of any dust explosion are influenced by or dependent upon the following factors:

1. The nature of the materials: Certain materials are more combustible, are more readily ignited, evidence greater pressure rises, and develop higher pressures than do others.
2. The size and shape of the particles: The smaller the particle size, the more readily the material can be ignited due to greater exposed surface area per unit weight. Also, the smaller the size, the more readily the particles can be suspended in air, creating an explosive mixture.
3. Concentration of the material in air: Generally, the smaller the particle size, the lower the explosive limit.
4. Ignition temperature.
5. Ease of ignition or energy required to initiate ignition of the dust cloud.
6. Rate of pressure rise of the material: The faster the pressure rise, the more difficult to vent, suppress, or control the dust explosion.
7. The maximum pressure developed from the explosion.

As indicated previously, there are many materials that evidence problems when in the finely divided state. These are too numerous to discuss individually, but the NFPA has developed standards and references (4) for a number of specific materials and operations commonly encountered in industry, and OSHA has a regulation, 29 CFR 1910.272 *Grain Handling Facilities* (5). These include starches, grains, flour and feed mills, plastics, sulfur, spices, confectionary, and woodworking. While these standards cover good loss prevention practices for many materials, the observer will note the common approach and protection when handling materials subject to dust explosions, as indicated earlier.

COMBUSTIBLE METALS

Some metals are highly combustible and present severe hazards when in a finely divided state; e.g., during machining, heat-treating, or grinding. Others, called *alkaline metals*, are strongly reactive with water and common extinguishing agents, and there are some that exhibit *pyrophoric* tendencies (the characteristic of igniting in air without an ignition source).

The water-reactive metals encountered are mainly sodium and potassium and their alloys, but occasionally one will encounter lithium, rubidium, and cesium. These materials, when brought in contact with water, will react and normally ignite. The use of dry chemical, CO_2, foam, or halogenated extinguishing agents should be avoided; most important, these metals should not be brought in contact with water, which causes the release of hydrogen. These metals are handled in hermatically sealed containers, and small quantities can be stored under a hydrocarbon such as kerosene.

Extinguishment of these metals is generally by dry inert materials, including sodium chloride and soda ash, except that these materials should not be used on lithium fires, which can be extinguished using graphite, lithium chloride powder, or zirconium silicates.

Metal hydrides, which are used as a source of hydrogen, are also reactive with water; as a consequence, common extinguishing agents cannot be applied. The hazards of the metal hydrides are similar to those of the alkaline metals.

Those involved in fire protection will more frequently encounter magnesium and, to a lesser extent, titanium and zirconium, all of which are highly combustible. Aluminum is not considered hazardous from a fire protection standpoint except when in a finely divided state or when used in nitrate salt baths. Good dust control design and loss prevention practices are essential to avoid explosions in aluminum powder facilities.

The hazard of magnesium depends upon its size and form; dust or shavings readily ignite. Heavy magnesium castings do not represent much of a hazard, unless involved in a large fire, because of the conductivity of the metal and the ease of dissipating heat. Because of the difficulty in combating fires involving large quantities of magnesium, it is necessary to give careful consideration to construction, which should be of a noncombustible or fire-resistant.

It is preferable to have magnesium operations in a building that does not require sprinkler protection, either from construction or from a combustible fuel loading, because water on magnesium fires will react violently and cause very intensive burning. In foundry areas where molten spills may occur, it is important that there be no moisture in pits or on the floor and that the floor materials contain no moisture such as might be found in ordinary concrete.

All magnesium operations are hazardous, specifically the casting, heat-treating, grinding, and machining operations. In the casting operations, pots should be cleaned of all iron scale and rust to prevent reaction of the iron oxide and the magnesium, which may form explosive thermite reactions. Pots should be checked frequently to ascertain that they are still sound. During a casting operation, flux composed of sulfur dioxide and boric acid is used on the molten magnesium surface, which seals and prevents combustion.

Heat-treating operations can be critical due to the relatively low autoignition point of magnesium. Dual excess-temperature limit controls are normally used, and the oven interior is provided with an atmosphere of sulfur dioxide to prevent ignition.

Grinding and machining operations present somewhat the same problems, with grinding being more hazardous because of the finely divided magnesium particles. This operation should only be conducted using wet-type dust collectors that confine the dust immediately by water spray and settle it in water in sludge form. It is important that the sludge be covered with liquid at all times because of the extremely hazardous nature of the dry material. Chambers of these wet collectors are provided with a 1-in. vent to dissipate the hydrogen that is evolved.

Magnesium machining operations produce accumulations of magnesium scrap that is readily ignitable in the shaving, ribbon, or small-particle form. The amount of magnesium must be kept to small quantities and scrap must not be allowed to accumulate, but should be removed to a safe location and disposed of safety. Also of considerable importance in preventing the ignition of magnesium during machining is the use of the proper tools, which should be kept sharp to prevent overheating, and also the use of mineral oil coolants with high flash points in place of conventionally used cutting oils.

Because titanium is pyrophoric, this material cannot be poured through the air. Unlike magnesium, water-type coolants are used for machining operations, using nonsparking hand tools for powdered materials. Small quantities of titanium can be extinguished by complete submersion in water.

Zirconium and certain other combustible materials, including uranium, are similar to magnesium and titanium. It should be kept in mind that all these materials can be very highly reactive and dangerous when encountered in industry.

PLASTICS

Until a relatively few years ago, the fire protection community, when evaluating the hazards of plastics, was satisfied with using such general classifications of hazard as "slow burning" and "moderate burning," based on small-scale tests. In recent years, plastic test fires conducted by the Factory Mutual Research Corporation (Factory Mutual Corner Tests) and loss experience have shown this approach to be completely inadequate. Tests and fire experience have shown that the results and ratings obtained by the commonly accepted test methods are in many cases invalid, particularly when applied to plastic foams such as polyure-

Table 6.5 Polymer Family

Thermosets	Thermoplastics	Elastomers
Alkyds	ABS	Butadiene
Allyls	Acetals	Butyl
Epoxies	Acrylics	Isoprene
Melamine	Cellulosics	Natural rubber
Polyesters	Nylons	Neoprene
Polyurethane	Olefins	Nitrile
Silicones	Polycarbonate	Urethane
Urea	Polystyrene	
	Vinyls	

thane and styrofoam, as well as to plastics such as styrenes. Presently, Factory Mutual Research Corporation, ASTM, and UL have developed a number of fire tests for plastics. In addition, the Society of the Plastics Industry has a number of Fact-Finding Reports and Bulletins that should be referenced.

Obviously, our dependency upon plastics is such that use of these materials cannot be avoided. Rather, it is a question of learning how to use and protect these materials in construction and other areas of our life more intelligently.

Plastics are actually synthetic resins with additives such as fillers, colorants, stabilizers, and plasticizers and are generally categorized in two classes: thermoplastics and thermosets. They are members of the family of polymers that are characterized by large molecules formed in chemical polymerization reactions. Also included in this family are elastomers, which are commonly referred to as synthetic rubbers (1). Some examples of the polymer family are shown in Table 6.5.

Generally, the thermoplastics present a greater hazard. The distinction between these two classes is the ability of thermoplastics to soften under heat with no chemical action taking place, with the material then hardened by cooling. Thermosetting plastics do not possess this characteristic and, when molded or formed, cannot be resoftened and reworked.

As stated earlier, traditional fire test methods have failed to predict the fire behavior of some plastics. Plastics in general have some special fire behavior problems. Some of these are:

Ignitability and burning rate: Some plastics are easily ignited and tend to have very high surface flame-spread rates.

Smoke: Some plastics are characterized by rapid generation of dense black smoke.

Toxic gases: Even though all fires generate toxic products of combustion, some plastics generate highly toxic gases such as hydrogen cyanide, hydrogen chloride, and phosgene.

Flaming drips: Flaming drips, which are characteristic of thermoplastics, can further propagate a fire by starting secondary fires.

Corrosion: Corrosion damage to metals and electronic equipment has been caused by fires in some plastics.

Fire hazards of cellular plastics have been a concern to the fire protection community, being used extensively throughout the construction industry primarily for insulation, but also in furniture in the form of upholstery materials.

In large quantities, these materials should be used with caution because of their flammability and low melting points. These materials, due to their relatively low cost and high insulating characteristics, will continue to be used in the future. Foam plastics are sold in preformed blocks or panels or, in the case of polyurethane, may also be sprayed or foamed in place. Fire-retardant grades are available that have a good resistance to temporary ignition sources, and the burning rate tends to be lower. Where these materials are found (even in sprinklered locations), building codes require a thermal barrier. This barrier should be a noncombustible covering such as gypsum board or other similar methods.

Plastics represent a varying degree of hazard depending upon the nature of the material, its use, and how it is stored. Plastic dust suspended in air is similar to other dusts with regard to fire and explosion, and the discussions on dust hazards preceding this section will also apply to plastics.

The result of large-scale tests conducted by the Factory Mutual Research Corporation show that most plastics stored in racks or in solid palletized form require large quantities of water properly applied to obtain control or extinguishment. Of considerable importance in evaluating the hazard, in addition to the nature of the plastic in high-piled storage, is a determination of whether there are numerous voids involved in the storage array that would expose the product to fire on several sides. An example of a high void factor would be storage of high-density molded parts on a plastic pallet with parts separated from each other by air spaces.

Major HPR underwriters will generally classify plastics into three or four categories, with the first or most hazardous classification consisting of the foam plastics polyurethane and styrofoam. The next category presenting a similar challenge to a protection system is polystyrene and acrylonitrile butadiene styrene (ABS). These materials degrade under fire conditions, with degradation taking the form of a depolymerization process where the polymer reverts to the monomer, which is a styrene monomer having a flash point approximately 90°F (32°C). Therefore, under fire conditions, these materials will exhibit fire characteristics similar to those of a flammable liquid and will have a British thermal unit release of approximately twice that of ordinary Class A combustible materials.

The third general classification includes polyethylene, polypropylene, and similar synthetic resins that evidence lesser flammable characteristics than do the styrenes. The fourth class includes polyvinyl chloride (depending upon the nature of the additives in the form of plasticizers or fillers) and various thermosetting plastics.

To summarize, in order to determine the density for ceiling sprinklers and the need for in-rack protection, it would be necessary to

1. Determine the nature of the plastic handled or to be handled
2. Determine the method of storage; that is, racks or palletized piling
3. Determine the storage configuration and whether or not the products to be handled will have void spaces as described above

If fire losses are to be prevented or controlled, it is necessary to start with a review of the hazards involved, which by necessity will require an in-depth knowledge of the flammability characteristics of the materials present. Knowing the hazards of the materials, whether liquid, gas, or solid, attention must be given to safeguarding the hazards presented, followed by determining the type and design of the protective systems that will be most effective for extinguishment or control. These protective systems will be covered in a subsequent chapter.

IDENTIFICATION SYSTEMS FOR HAZARDOUS MATERIALS

The need to identify the flammable characteristics of materials stored in drums, tanks, tank cars, and other containers, vessels, or areas has long been recognized. This need has led to the use of a multitude of identification systems, using color with a pictorial view, a pictorial layout plus the use of English language descriptions, a pictorial display with numerical designations, or numerical designations only. While many attempts have been made to develop a common system for labeling and placarding, none has been universally adopted, resulting in a number of different systems in use today by the U.S. DOT, the United Nations (UN), the NFPA, individual industrial firms, and others.

The most common systems in use today are the DOT regulations (12) and the NFPA's *Standard System for the Identification of the Fire Hazards of Materials*, NFPA 704 (4). The UN system for identification of hazardous materials will also affect some managers involved in international operations, so this system is also briefly reviewed.

For those involved in shipment of hazardous materials, it is essential that they become thoroughly familiar with the Hazardous Materials Regulations of the Materials Transportation Bureau, U.S. DOT (12). As in past years, DOT continues to use the familiar diamond-shaped label with various colors plus pictorial displays and English language descriptions to identify hazards. A few of the more commonly used labels and placards and those of greater interest to the average individual involved with hazardous materials are shown in Fig. 6.9a–c.

The UN system is similar to that employed by DOT, but also uses a numerical hazard class designation that is attached to the lower portion of the diamond.

The type of label to be used for hazardous materials in shipment is given in the DOT Hazardous Materials Table located in the previously mentioned DOT regulations. The table, in addition to listing the hazard class and label, also lists maximum quantities allowable in one package for shipment by air and water.

Both the NFPA and DOT systems use red coloring to identify a flammable material, with the DOT system also using red and white stripes to indicate a flammable solid and red lower half and white upper half to designate a spontaneously combustible material. Yellow is used to indicate an oxidizing material by both NFPA and DOT. Other colors used by DOT are orange for explosives, green for nonflammable gases, white for poisons, and blue for "dangerous when wet" materials.

Returning to the NFPA system for identification of hazards of materials, Figure 6.10 shows the diamond-spatial arrangement divided into four separate smaller diamonds, each with a different color. The top diamond in red with a numeral from 0 to 4 identifies the flammability. The blue diamond to the left indicates the health hazard with a similar numerical system. The right diamond of yellow, also with one of five numerals (0 through 4), identifies reactivity or stability. The fourth or bottom space of white is used primarily to identify unusual reactivity, with water denoted by the letter W with a line through the center, $\overline{W}$. The lower diamond can also be used to identify other possible problems, such as radioactivity, an oxiding agent, or the proper extinguishing agent.

The basic purpose of any identification system is to alert handlers of hazardous materials and fire-fighting personnel of the danger involved in the inherent nature of materials confined in packages or other enclosed containers. With this in mind, the NFPA system accomplishes this objective to a much greater degree. The table also shown in Figure 6.10 gives the significance of the various numerical designations of the NFPA system for health hazards, flammability, and reactivity. A number of communities have adopted the use of this system. For instance, in Boston, Mass., this identification system is required on all laboratory doors of research installations to alert fire-fighters.

DOT Classifications	United Nations class
Class A explosives	1
Class B explosives	1
Class C explosives	1
Flammable compressed gas	2
Nonflammable compressed gas	2
Flammable liquid	3
Flammable solid	4
Oxidizing material	5
Poisonous gas, Class A	2 or 6
Poisonous liquid, Class A	6
Poisonous liquid or solid, Class B	6
Irritating material	6
Etiologic agent	6
Radioactive materials	7
Corrosive material	8

Figure 6.9   Department of Transportation hazardous materials warning labels and placards. (Courtesy of Department of Transportation.)

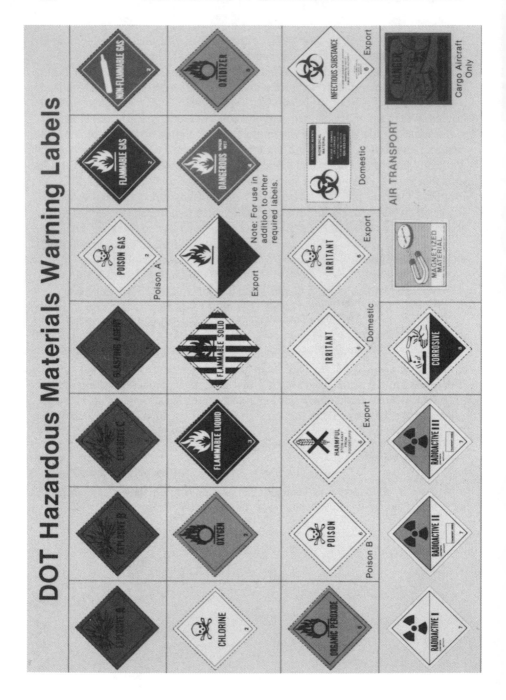

General Guidelines on Use of Labels

1. The Hazardous Materials Tables, Sec. 172.101 and 172.102, identify the proper label(s) for the hazardous materials listed.

2. Any person who offers a hazardous material for shipment *must label* the package, if required. [Sec. 172.400(a)]

3. Labels *may* be affixed to packages (even though not required by the regulations) provided *each* label represents a hazard of the material in the package. [Sec. 172.401]

4. Label(s), when required, *must* be printed on or affixed to the surface of the package near the proper shipping name. [Sec. 172.406(a)]

5. When two or more different labels are required, display them next to each other. [Sec. 172.406(c)]

6. When two or more packages containing compatible hazardous materials are packaged within the same overpack, the outside container *must* be labeled as required for each class of material contained therein. [Sec. 172.404(b)]

7. Material classed as an **Explosive A, Poison A,** or **Radioactive Material** also meeting the definition of another hazard class *must* be labeled for *each* class. [Sec. 172.402(a)]

8. Material classed as an **Oxidizer, Corrosive, Flammable Solid,** or **Flammable Liquic** that also meets the definition of a Poison B *must* be labeled POISON, in addition to the hazard class label. [Sec. 172.402(a)(3) and (5)]

9. Material classed as a **Flammable Solid** that also meets the definition of a water-reactive material *must* be labeled with FLAMMABLE SOLID and DANGEROUS WHEN WET labels. [Sec. 172.402(a)(4)]

10. Material classed as a **Poison B, Flammable Liquid, Flammable Solid,** or **Oxidizer** that also meets the definition of a Corrosive material *must* be labeled CORROSIVE in addition to the class label. [Sec. 172.402(a)(6) through (9)]

Hazardous Materials Class Numbers

Hazardous materials class numbers associated with the hazard classes.

Class 1- Explosives
Class 2- Gases (Compressed, liquefied or dissolved under pressure)
Class 3- Flammable liquids
Class 4- Flammable solids or Substances
Class 5- Oxidizing Substances
Class 6- Poisonous and Infectious Substances
Class 7- Radioactive Substances
Class 8- Corrosives
Class 9- Miscellaneous dangerous Substances

NOTE: For requirements, see Sec. 172.102(h), 172.332(c)(3) and 172.407(g).

This chart is designed as a reference. For more complete details, refer to the *Code of Federal Regulations*, Title 49, Parts 100-177

US Department of Transportation

Research and Special Programs Administration

Materials Transportation Bureau
Washington, D.C. 20590

Chart 7 September 1982
Revised

Figure 6.9 Continued

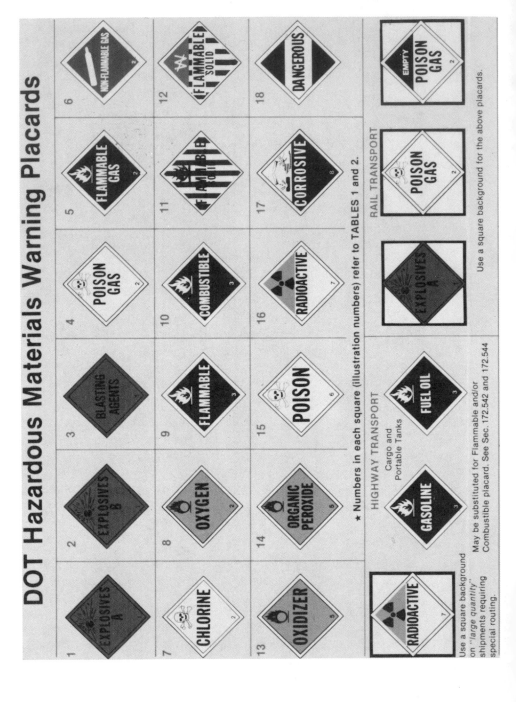

DOT Hazardous Materials Warning Placards

★ Numbers in each square (illustration numbers) refer to TABLES 1 and 2.

Use a square background on "large quantity" shipments requiring special routing.

May be substituted for Flammable and/or Combustible placard. See Sec. 172.542 and 172.544

Use a square background for the above placards.

HIGHWAY TRANSPORT
Cargo and Portable Tanks

RAIL TRANSPORT

TABLE 1

Hazard Classes	★ No.
Class A explosives	1
Class B explosives	2
Poison A	4
Flammable solid (DANGEROUS WHEN WET label only)	12
Radioactive material (YELLOW III label)	16
Radioactive material:	
Uranium hexafluoride, fissile (containing more than 0.7% U235	16 & 17
Uranium hexafluoride, low-specific activity (containing 0.7% or less U235	16 & 17

TABLE 2

Hazard Classes	★ No.
Class C explosives	18
Blasting agent	3
Nonflammable gas	6
Nonflammable gas (Chlorine)	7
Nonflammable gas (Fluorine)	15
Nonflammable gas (Oxygen, pressurized liquid)	8
Flammable gas	5
Combustible liquid	10
Flammable liquid	9
Flammable solid	11
Oxidizer	13
Organic peroxide	14
Poison B	15
Corrosive material	17
Irritating material	18

Guidelines

- Placard motor vehicles, freight containers, and rail cars containing any *quantity* of hazardous materials listed in TABLE 1.
- Placard motor vehicles and freight containers containing 1,000 pounds or more gross weight of hazardous materials classes listed in TABLE 2.
- Placard freight containers 640 cubic feet or more containing any quantity of hazardous materials classes listed in TABLES 1 and/or 2 when offered for transportation by air or water. Under 640 cubic feet, see Sec. 172.512(b).
- Placard rail cars containing *any quantity* of hazardous materials classes listed in TABLE 2 except when less than 1,000 pounds gross weight of hazardous materials are transported in TOFC (Trailer on Flat Car) or COFC (Container on Flat Car) service.

UN and NA Identification Numbers

1. UN (United Nations) or NA (North American) numbers are found in Sec. 172.101, 172.102 and the Emergency Response Guidebook.

2. The four-digit UN or NA numbers are used to identify the hazardous materials involved.

3. NA numbers are used only in the USA and Canada.

4. UN or NA numbers must be displayed on Tank Cars, Cargo Tanks and Portable Tanks.

5. When ID numbers are displayed on placards, ORANGE PANELS are not required.

6. When ID numbers are displayed on ORANGE PANELS, appropriate placards are also required.

7. EUROPEAN NUMBERING SYSTEM. Top numbers represent the Hazard Index. The bottom numbers are the required UN identification numbers.

For more compete details on Identification Numbers see Sec. 172.300 through 172.338.

Figure 6.9 Continued

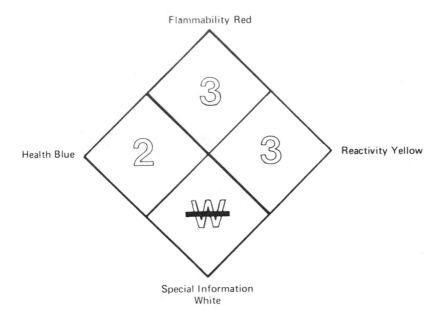

Figure 6.10 National Fire Protection Association's system for idenification of hazards of materials (NFPA 704 diamond). (Reprinted with permission from NFPA704–1985, *Identification of the Fire Hazards of Materials*, Copyright 1985, National Fire Protection Association, Quincy, Mass. This reprinted material is not the complete and official position of the NFPA on the referenced subject, which is represented only by the standard in its entirety.)

Identification of Health Hazard Color Code: BLUE	Identification of Flammability Color Code: RED	Identification of Reactivity (Stability) Color Code: YELLOW
Type of Possible Injury	Susceptibility of Materials to Burning	Susceptibility to Release of Energy
Signal	Signal	Signal
4 — Materials which on very short exposure could cause death or major residual injury even though prompt medical treatment were given.	4 — Materials which will rapidly or completely vaporize at atmospheric pressure and normal ambient temperature, or which are readily dispersed in air and which will burn readily.	4 — Materials which in themselves are readily capable of detonation or of explosive decomposition or reaction at normal temperatures and pressures.
3 — Materials which on short exposure could cause serious temporary or residual injury even though prompt medical treatment is given.	3 — Liquids and solids that can be ignited under almost all ambient temperature conditions.	3 — Materials which in themselves are capable of detonation or explosive reaction but require a strong initiating source or which must be heated under confinement before initiation or which react explosively with water.
2 — Materials which on intense or continued exposure could cause temporary incapacitation or possible residual injury unless prompt medical treatment is given.	2 — Materials that must be moderately heated or exposed to relatively high ambient temperatures before ignition can occur.	2 — Materials which in themselves are normally unstable and readily undergo violent chemical change but do not detonate. Also materials which may react violently with water or which may form potentially explosive mixtures with water.
1 — Materials which on exposure would cause irritation but only minor residual injury even if no treatment is given.	1 — Materials that must be preheated before ignition can occur.	1 — Materials which in themselves are normally stable, but which can become unstable at elevated temperatures and pressures or which may react with water with some release of energy but not violently.
0 — Materials which on exposure under fire conditions would offer no hazard beyond that of ordinary combustible material	0 — Materials that will not burn.	0 — Materials which in themselves are normally stable, even under fire exposure conditions, and which are not reactive with water.

Figure 6.10 Continued

REFERENCES

1. Cote, Arthur E., ed., *Fire Protection Handbook*, 16th Edition, National Fire Protection Association, Quincy, Mass., 1986.
2. *Loss Prevention Data Books*, Factory Mutual Engineering Corporation, Norwood, Mass., 1989.
3. *Flash Point Index of Trade Name Liquids*, 9th Edition, National Fire Protection Association, Quincy, Mass., 1978.
4. *National Fire Codes*, National Fire Protection Association, Quincy, Mass., 1989.
5. *General Industry Standards—29 CFR 1910*, Occupational Safety and Health Administration, Washington, D.C., 1988.
6. *Approval Guide*, Factory Mutual Engineering Corporation, Norwood, Mass., 1989.
7. *Venting Atmospheres and Low Pressure Storage Tanks*, API Standard 2000, 2nd Edition, American Petroleum Institute, Washington, D.C., 1973.
8. Linville, Jim L., ed., *Industrial Hazards Handbook*, 2nd Edition, National Fire Protection Association, Quincy, Mass., 1984.
9. "Safe Handling of Compressed Gases in Containers," Pamphlet P–1, Compressed Gas Association, New York.
10. Compressed Gas Association, *Handbook of Compressed Gases*, Van Nostrand Reinhold Co., New York.
11. *Matheson Gas Data Book*, The Matheson Co., Inc., East Rutherford, N.J.
12. *Code of Federal Regulations*, Title 46, Parts 146–149, Title 49, Parts 171–190, Department of Transportation, Washington, D.C., 1989.
13. *ASME Boiler and Pressure Vessel Code*, American Society of Mechanical Engineers, New York, 1977.
14. Bochnak, Peter M., Pizatella, Timothy, J., and Lark, Joseph J., "LP-Gas Emergencies—A Review and Appraisal of Selected Ignited Leaks During Liquid Transfer Operations," National Institute for Occupational Safety and Health, Morgantown, W. Va., 1981.
15. *The Condensed Chemical Dictionary*, 10th Edition, Van Nostrand Reinhold Co., New York, 1981.
16. Sax, N. Irving and Lewis, Sr., Richard J., *Hazardous Chemicals Desk Reference*, Van Nostrand Reinhold Co., New York, 1987.

7

Common Process Hazards

This chapter discusses some of the hazards involved in common processes and operations that find widespread use in the industrial setting. Of course, each facility will have its own unique processes or operations. As stated in Chapter 3, Elements of a Fire Loss Control Program, the hazards in all processes specific to each facility must be identified and evaluated to establishing priorities for the concentration of efforts in the loss control program.

SPRAY BOOTHS: DESIGN AND PROTECTION

A common operation extensively conducted in industry and involving the use of flammable liquids is spray finishing. This operation is covered in detail by the *Standard For Spray Application Using Flammable and Combustible Materials*, NFPA 33,1 and 29CFR1910.107, *Spray Finishing Using Flammable and Combustible Materials* 2.

Ideally, this operation should be conducted in detached one-story building or in one-story portions of buildings, adequately separated from other important operations by fire walls. This is not always possible in modern plants due to product flow and conveyor lines. Assuming automatic sprinkler protection, which should be considered basic for flammable liquid protection and confining quantities of flammable liquids outside of storage rooms to a minimum quantity, the use of draft curtains, preferably with roof venting, is considered an acceptable alternative to cutoffs in many situations.

The use of basements for these operations or for any flammable liquid processing should be prohibited. It is also not desirable to conduct finishing operations in upper stories of multistory buildings. Where they must be conducted in upper stories, waterproofing of floors and curbing along with reduction of high values immediately below these points are items to consider.

Of interest is a section of NFPA 33, which states, "Industrial and similar business buildings having spray finishing operations should be so located and protected as to minimize possible damage to other property by fire or extinguishing agents" (1). This recognizes the fact that it is often impossible to provide a desirable cutoff, but considerable thought should be given to safely locating the hazard with regard to exposure of adjacent high values. At the same time, this standard calls for vertical and horizontal separation of spray-finishing operations in building classified as assembly, educational, institutional, or residential by noncombustible construction having a minimum 2 hr of fire resistance.

Spray-finishing operations may also be conducted in rooms, which would define the entire spray area as the entire room enclosure in the event of nonuse of booths. Assuming the storage of flammable liquids in properly designed storage rooms or buildings, a minimum fire-resistance rating of 1 hr is generally adequate for a spraying enclosure provided with sprinkler protection.

Spray booths are characterized in two general categories including the dry and water-wash types. Water-wash types are considered preferable to dry types because they collect and reduce the hazard of overspray by limiting the quantity of overspray entering the exhaust ducts and discharged from the ventilating system. A typical water-wash booth is shown in Figure 7.1. Booths should be designed with minimum 18-gauge steel with noncombustible floors or, if combustible, covered with a noncombustible material to facilitate safe cleaning and removal of residue. A metal deflector curtain not less than 2.5 in. deep at the upper outer edge of the booth should be provided if the booth has a frontal area larger than 9 ft^2. It is important that spray booths be constructed and installed so that all portions are readily accessible for cleaning.

Spray booths having dry-type filters or filter rolls should be designed, installed, and maintained so that the average air velocity across the face of the booth will not be less than 100 lineal ft/min except for electrostatic spraying operations, where the air velocity should not be less than 60 lineal ft/min. Visible gauges or audible alarm or pressure-activated devices should be provided to indicate that the required air velocity is available.

All spraying areas are required to have mechanical ventilation to remove flammable vapors to a safe location. For air-operated guns manually or automatically operated in booths with cross drafts of up to 50 ft/min, a design airflow velocity of 100 ft/min should be used for large booths, with a range of from 75 to 120 ft/min. For small booths, the design airflow velocity should be 150 ft/min, with a range of

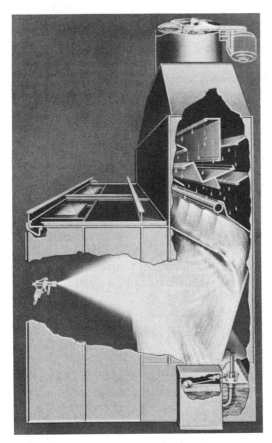

Figure 7.1 A water-wash type booth used in spray painting. (Courtesy of DeVil-biss Co.)

125 to 170 ft/min. For cross drafts of up to 100 ft/min, the design airflow veloci-ties 150 ft/min for large booths and 200 ft/min for small booths. Note that NFPA 33 requires 100 lineal ft/min, except for electrostatic spray, where 60 ft/min is acceptable (1).

Electrostatic spray-finishing operations conducted in a booth should have a minimum air velocity of 50 ft/min for large booths, with a design range of 50 to 75 ft/min. For small booths, the minimum airflow velocity should be 100 ft/min, with a range of 75 to 125 ft/min. This reduction in air velocity recognizes that there is considerably less overspray involved in electrostatic spraying operations. In addition to the above, the total air volume exhausted through a spray booth

Table 7.1   Lower Explosive Limits and Cubic Feet of Vapor per Gallon of the Liquid at 70°F (21°C) for Some Commonly Used Flammable Liquids

Solvent	LEL in percent by volume of air at 70°F	Cubic feet of vapor per gallon of liquid at 70°F[a]
Acetone	2.6	44.0
Benzene	1.4	36.8
Cellosolve acetate	1.7	23.2
Ethyl acetate	2.5	32.8
Ethyl alcohol	4.3	55.2
Methyl alcohol	7.3	80.8
Methyl ethyl ketone	1.8	36.0
Naphtha, VM and P	0.9	22.4
Toluene (toluol)	1.4	30.4
Xylene (xylol)	1.0	26.4

[a]The cubic feet of vapor per gallon of liquid that will render an equivalent quantity of air barely explosive.

should be sufficient to dilute solvent vapors to at least 25% of the lower explosive limit of the solvent being sprayed. This airflow can be determined using the following formula:

Dilution volume required per gallon of solvent =

$$\frac{4(100 - LEL) \text{ cubic feet of vapor per gallon}}{LEL}$$

Table 7.1 gives LEL and cubic feet of vapor per gallon of the liquid at 70°F (21°C) for some commonly used flammable liquids.

The dilution volume required for a specific situation is obtained by multiplying the dilution volume required per gallon of solvent by the number of gallons of solvent sprayed per minute to obtain the cubic feet per minute of fan capacity.

When using average air velocities, the capacity in cubic feet per minute of the exhaust fan is obtained by multiplying the square footage of the open frontal area of the booth by the average velocity in linear feet per minute. Mechanical ventilation should be kept in operation at all times while spraying and for an adequate time thereafter to remove all vapors. Wherever practical, the air supply should be interlocked with the ventilating system to prevent spraying unless the exhaust system is in operation.

The following further ventilating provisions are important:

Fan blades should be nonferrous or nonsparking where the casing consists of or is lined with such material.

Electrical motors, although of Class I, Division 1, should not be installed in any spray area where they may be subject to deposits of overspray. This precludes the installation of motors inside booths or ducts.

Belts should not enter duct or booth unless the belt and pulley within the duct or booth are thoroughly enclosed.

Exhaust ducts should have a clearance of not less than 18 in. from unprotected combustible construction or other combustible material, and should be the minimum gauge required by NFPA 33 (1).

Discharge from a dry-type booth should not be less than 6 ft from any combustible exterior wall or roof, nor discharged in a direction of any combustible construction or unprotected opening in any combustible exterior wall within 25 ft.

Air exhausted from spray operations should not contaminate makeup air being introduced into the spraying area, nor should this air be recirculated.

Ventilation for enclosed spray rooms should, as a minimum, be 10,000 ft^3 of air at 70°F (21°C) for each gallon of solvent evaporated (with the maximum rate of application) with floor level pickup.

Makeup air should be provided to ensure good air movement and should preferably be fresh air directly from outdoors.

All sources of ignition should be prohibited within 20 ft of spraying area unless separated by a partition. Steam pipes or similar hot surfaces should not be located in the spraying area where deposits of combustible residue may readily accumulate.

For control of electrical ignition sources in spray areas, electrical wiring and equipment should be approved for Class I or Class II, Division 1 locations. The electrical wiring and equipment located adjacent to the immediate spray area and within 20 ft horizontally and 10 ft vertically should be approved for Class I or Class II, Division 2 locations (see Fig. 7.2). Reference should be made to NFPA 33 for other configurations for areas adjacent to the spray area (1).

Protection against static accumulation is necessary in spray finishing. This is accomplished by properly grounding all metal parts of spray booths, exhaust ducts, and any piping systems conveying flammable liquids.

Standard sprinkler protection is desirable for all spray-finishing operations, installed preferably as a hydraulically designed system furnishing a density of 0.30 gal/min per square foot or installed on an "extra hazard" schedule. Systems should be wet-pipe. Sprinkler heads in the booth should be located so that they are not subject to overspray, but may be covered with thin paper or plastic bags to prevent buildup of deposits. Sprinklers protecting the spray booth, together with heads in connecting exhaust ducts, should be controlled by a separate OS & Y valve that is readily accessible. While sprinklers protecting ducts should prefer-

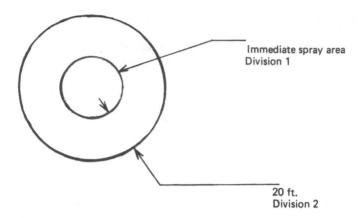

Immediate spray area
Division 1

20 ft.
Division 2

Figure 7.2 Class I or Class II, Division 1 and 2 locations adjacent to an unenclosed spray area.

ably be of a closed-head automatic type on a nonfreeze system, where required, open heads controlled manually are permissible, but less desirable. Sprinklers in horizontal ducts should be installed not more than 12 ft apart and should be accessible through access doors in the ducts for replacement as needed.

DIP TANKS: DESIGN AND PROTECTION

Dip-tank installations containing flammable liquids require close adherence to good loss control practices to avoid serious fires and even explosions. The seriousness of any dip-tank fire will depend upon the nature and quantity of the flammable liquid contained within the tank, the location, and the protection or lack of protection. This operation is covered in detail by the *Standard for Dipping and Coating Processes Using Flammable and Combustible Liquids*, NFPA 34 (1) and 29CFR 1910.108, *Dip Tanks Containing Flammable or Combustible Liquids* (2). Such operations should preferably be located in detached one-story buildings or, when located in industrial and similar buildings, should be separated from other important occupancies by noncombustible construction having a minimum 2-hr fire-resistance rating. Since this is not always practical from a standpoint of product flow, good fire protection practices are essential for proper safeguarding. Figure 7.3 shows an example of a typical dip-tank process system.

In lieu of isolation or separation by partitions of a suitable fire-resistance rating, draft curtains may be acceptable when installed surrounding the hazard and used in conjunction with automatic smoke and heat venting. Dipping operations should not expose concentrations of high value or be located in close proximity to

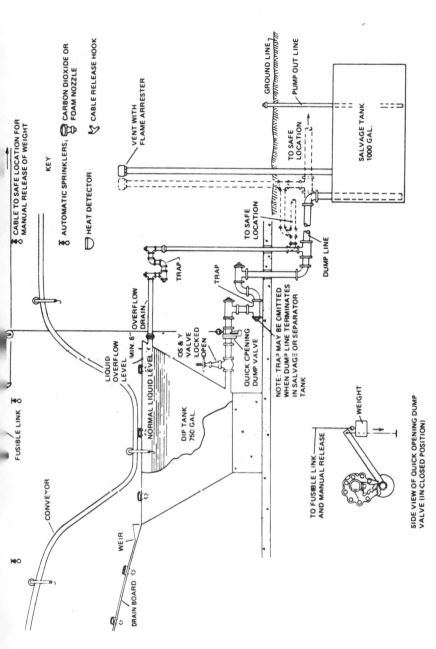

KEY

CABLE TO SAFE LOCATION FOR MANUAL RELEASE OF WEIGHT

AUTOMATIC SPRINKLERS

CARBON DIOXIDE OR FOAM NOZZLE

HEAT DETECTOR

CABLE RELEASE HOOK

VENT WITH FLAME ARRESTER

GROUND LINE

PUMP OUT LINE

TO SAFE LOCATION

SALVAGE TANK 1000 GAL.

TO SAFE LOCATION

DUMP LINE

TRAP

TRAP

OVERFLOW DRAIN

MIN. 6"

LIQUID OVERFLOW LEVEL

NORMAL LIQUID LEVEL

OS & Y VALVE LOCKED OPEN

DIP TANK 750 GAL.

QUICK OPENING DUMP VALVE

NOTE: TRAP MAY BE OMITTED WHEN DUMP LINE TERMINATES IN SALVAGE OR SEPARATOR TANK

CONVEYOR

WEIR

DRAIN BOARD

FUSIBLE LINK

WEIGHT

TO FUSBLE LINK AND MANUAL RELEASE

SIDE VIEW OF QUICK OPENING DUMP VALVE (IN CLOSED POSITION)

Figure 7.3 Example of a dip-tank process system. Reprinted with permission from the *Fire Protection Handbook*, 16th Edition, Copyright 1986, National Fire Protection Association, Quincy, Mass.

processing or utilities essential to the continuity of operations. Curbing and drainage around dip tanks may be advisable to confine any possible flammable liquid spill. Storage should obviously be in conformance with recommendations set forth in Chapter 6, Fire Hazards of Materials and Their Control.

All dip tanks and drain boards should be constructed of noncombustible material with supports of heavy metal-reinforced concrete or masonry. On occasion, dip tanks may be found set into a floor extending to the story below. In such situations, it is advisable to protect any supports to provide a minimum 1-hr fire-resistance rating. In addition, to prevent water flowing into the tank, the top of the dip tank should be not less than 6 in. above the floor of the room. Opening around the dip tank between the tank and the floor should be sealed to provide a fire-resistance rating equal to that of the floor.In multistory buildings, it is preferable that dip tanks be located on first levels. Basement locations should be prohibited.

Dip tanks of over 150 gal in capacity and 10 ft^2 in liquid surface area should be provided with a rapped overflow pipe leading to a safe location outside the building. The purpose of the overflow is to provide for the discharge of overflow caused by the application of water during fire fighting. This requires an overflow pipe of sufficient size to remove the maximum rate of flow of water expected to be applied from automatic sprinklers or other fire suppression means. NFPA 34 specifies a minimum overflow pipe of 3 in. in diameter, with 4 in. required for liquid surface areas of dip tanks of 75 to 150 ft^2, a diameter of not less than 5 in. for 150 to 275 ft^2 of liquid area tanks, and not less than 6 in. in diameter for tanks having a liquid surface area of 225 to 325 ft^2 (1). The bottom of the overflow connection should not be less than 6 in. below the top of the tank and should be connected to the tank through a flared outlet where any accumulation of caked or dried material could clog the overflow opening.

Emergency bottom outlets should be installed on dip tanks over 500 gal in liquid capacity arranged to drain both automatically and manually in the event of fire to reduce the intensity and duration. These outlet drains are normally weighted valves actuated by fusible links, draining either to a safe exterior location or to a salvage tank that has a greater capacity than the dip tank to which they are connected. All drains to the tanks, including the overflow drains, should be properly trapped to prevent passage of vapors. Emergency drains are sized in accordance with tank capacity, as specified in NFPA 34 with 3-in.-diameter drains required for tanks of 500-to 750-gal capacity, 4 in. for 750-to 1000-gal. capacity, 5 in. for 1000-to 2500-gal. 6 in. for 2500-to 4000-gal capacity, and 8 in. for tanks in excess of 4000-gal liquid capacity (1).

Ventilation by mechanical means should be provided to confine the vapor area to the smallest practical volume. This may be accomplished by installation of local slot-type ventilation attached to the tank side or by general room ventilation with low-level pickup.

The amount of ventilation should be sufficient to restrict flammable vapor concentrations to a maximum of 25% of the lower explosive limit within 2 ft of the dip tank or drain board. Generally, with room ventilation, this will require 1 ft³/min per square foot of floor area. It should be noted that ventilation required for employee health will normally be greater than that required for confinement of vapors to well below the lower explosive limit.

No ignition sources should be permitted in any vapor area. Electrical installations should conform to the National Electric Code, NFPA 70, Article 500, and within any vapor area should be of Class I, Division 1 type (1). This requirement will apply to distances within 10 ft horizontally from the dip tank and drain board to a height of 3 ft above the highest point of vapor liberation.

Dip-tank areas should be protected by a system of automatic sprinklers, and all dip tanks exceeding 150-gal liquid capacity or having a liquid surface area exceeding 4 ft² should be protected by at least one of the following systems:

An automatically closing dip-tank cover
An automatic carbon dioxide extinguishing system
An automatic dry-chemical extinguishing system
An automatic Halon 1301 or 1211 extinguishing system
An automatic water-spray extinguishing system

Dip-tank covers are obviously impractical except for small tanks, which suggests the use of an automatic fixed extinguishing system for protection.

WELDING AND CUTTING OPERATIONS AND OTHER HOT WORK

Facility modifications and equipment repair frequently require welding and cutting. These operations are potentially very hazardous. As stated in the section on "Permit Systems" in Chapter 3, cutting and welding account for about 6% of the fires reported in industrial properties (1). Welding and cutting operations are considered hot work, along with brazing, soldering, grinding, using explosive-actuated tools, and any other operations that could produce high amounts of heat or energy. Hot work has accounted for more fires and explosions in facilities for example, in grain-handling facilities than has any other known cause (3). The extremely high temperatures and sparks generated during welding and cutting operations dictate the need for strict controls. As stated under "Permit Systems," use of a permit system is one effective control measure. Permits are needed for all hot work performed outside designated maintenance areas to ensure that necessary precautions have been taken.

The permit provides the written authorization, of a supervisor or other qualified person, to perform the work. It is signed only after the work site has been inspected and it has been verified that the necessary precautions have been taken. The permit is also signed by the persons performing the work and by support personnel to indicate that they are aware of the potential hazards and safe work practices that should be followed. It is particularly important that contractors follow the permit requirements, since they may not be familiar with the fire and explosion hazards in the facility. Prior to issuing a permit, the supervisor or qualified person should determine whether the work can reasonable be moved to a designated maintenance area or a nonhazardous area outside the facility. Alternative methods such as the use of hand saws or bolt fasteners may also minimize or eliminate the need for hot work. Although these alternative techniques are not always practical, they should be considered and evaluated before issuing the permit (3).

Some sources imply that a permit should be issued for each specific job. Others indicate that the permit should be renewed daily or at the beginning of each shift. Since a major purpose of the permit is to verify that the operator is familiar with the hazards and the safety precautions, it is recommended that the permit be renewed at the beginning of each shift.

Employees performing welding, cutting, or other hot work should be properly instructed and qualified to operate the equipment and should be made aware of the fire hazards. Outside contractors should be instructed on the specific fire and explosion hazards that they may encounter in the facility.

Cutting or welding operations should not be permitted in (2):

Areas not authorized by management
Sprinklered buildings while protection is *impaired*
The presence of explosive atmospheres (mixtures of flammable gasses, vapors, liquids, or dusts with air)
The presence of explosive atmospheres that may develop inside uncleaned or improperly prepared tanks or equipment
Areas of large quantities of exposed, readily combustible materials

Cutting or welding should be permitted only in areas that are made fire safe. For example, special precautions are necessary when there is an exposure to an area that has an atmosphere containing flour, starch, or grain dust. Complete shutdown of the facility before conducting any hot work in these areas is recommended. Where the entire facility cannot reasonable be shut down, dust-producing operations must be terminated within the work areas and in adjacent areas where airborne dust could reach the work area. Necessary precautions, such as lockout techniques, should be taken to prevent inadvertent start-up of equipment while it is being worked on or where airborne dust could be produced. Equipment should remain off until the hot work has been completed and cooled, the area has been inspected for residual heat and smoldering fires, and the equipment has been approved for restart (3).

Combustible materials within 35 ft of the work area should be removed. When materials cannot reasonably be removed, they should be protected by fire-resistant shields or covers. Wetting of combustible materials in the area is recommended as an additional precaution. Care should be taken to protect combustibles that may be concealed from view. Floors, ledges, and other surfaces within 35 ft of the hot-work area should be thoroughly cleaned of combustible debris. When hot work is elevated, the area underneath, taking into account for the wind drift of the hot slag and sparks, should also be cleaned. Where hot work is performed near walls or floors, adjacent areas should also be inspected and cleaned. Wall, floor, and other openings must be sealed where sparks or hot slag may reach.

A standby person (fire watcher) with a fire extinguisher or fire-extinguishing equipment is needed to monitor the area while the hot work is being performed and for at least 1/2 hr after cessation of the hot work. Additional checks up to 2 hr or more are recommended. A thorough inspection of the work area and adjacent areas should be made for residual heat and smoldering fires before the standby person leaves. If a security guard is employed during nonoperating hours, he or she should be advised that hot work has taken place.

Welding, cutting, and brazing equipment should be used in accordance with the manufacturers' instructions. Operations and equipment should comply with 29 CFR 1910, Subpart Q, *Welding, Cutting, and Brazing* 2; and for more detailed information, reference should be made to the *Standard for the Design and Installation of Oxygen-Fuel Gas Systems for Welding, Cutting, and Allied Processes*, NFPA 51, and *Standards for Fire Protection in Use of Cutting and Welding Processes*, NFPA 51B (1).

REFERENCES

1. *National Fire Codes*, National Fire Protection Association, Quincy, Mass., 1989.
2. *General Industry Standards—29 CFR 1910*, Occupational Safety and Health Administration, Washington, D. C., 1988.
3. *Occupational Safety in Grain Elevators and Feed Mills*, U. S. Department of Health and Human Services, Morgantown, W. Va., 1983.

8

Fire Safety in Design and Construction

This chapter will outline engineering design considerations for new plant projects or expansion as related to specific areas of fire protection and suppression. The discussion will include plant site selection; plant layout, with emphasis on accessibility for suppression and isolation of hazardous operations; the use of various construction materials; construction-related items such as smoke and heat venting; explosion-relief venting; building subdivision and blast-resistant enclosures; electrical installations; heating, ventilating, and air conditioning; and dust-collecting systems. Water supplies and automatic detection and suppression systems will be included in subsequent chapters.

Before discussing each specific area, it should be emphasized that fire protection input to plant design should be provided as early as possible and preferably at preengineering conferences. Although fire protection generally represents a small portion of the overall project cost, this expense can be substantial. Early involvement of the fire protection engineer can possibly avoid the necessity of costly change orders or addendums. While the fire protection engineer is, or should be, cost conscious, factors in addition to cost-benefit ratios and cost effectiveness must be included in the decision-making process to determine whether or not to provide certain levels of protection or structural features. One major consideration is the uniqueness of a facility or operation and the importance of maintaining continuity of operations of a profit center. It is suggested that this type of consideration is often of greater importance than a cost-benefit ratio or immediate cost effectiveness.

As part of this and subsequent sections, various protection systems will be described along with other features of construction or equipment related to fire protection in general.

SYSTEMS APPROACH

Fire safety can be incorporated into facility design in several different ways. One way is to require that the building be designed strictly to specification codes, such as building codes or military and government specifications. These are normally very stringent and have very little flexibility. They often do not take into account overall facility designs and often need to be interpreted because of confusion in certain specifications and lack of coverage of certain design features.

Another way to incorporate fire safety is to use performance codes. These codes try to overcome the inflexibility of specification codes by defining the expected fire safety performance of a separate component of the design and allowing alternative ways to meet the design. This approach has problems when one looks at the building as a whole design, since some components have a better chance of successful performance than do others.

The final approach is to look at fire safety as an integrated subsystem of the building, along with the functional, structural, electrical, or mechanical subsystems. This method relies on the use of the best engineering methodology rather than on strict compliance with codes. In other words, this approach provides and "equivalent" alternative to code requirements that provide equivalent fire safety. In the long run, this is the best way to incorporate fire safety into the overall design.

To effectively incorporate the facility's fire protection considerations into the design, the fire safety objectives must be identified. One descriptive tool to help designers identify the fire safety objectives in a systems approach is the systems tree of fire protection. This concept was first developed by the NFPA in its "Fire Safety Concepts Tree" (*Guide to the Fire Safety Concepts Tree*, NFPA 550) and by the General Services Administration in its system tree (1, 2, 3). A trimmed-down version developed by the Architecture Life Safety Group, University of California at Berkeley, for the U.S. Fire Administration, Department of Commerce, is shown in Figure 8.1 (4). Reference should also be made to the *Guide to the Fire Safety Concepts Tree*, NFPA 550, for a more detailed fire safety concepts tree and explanation and examples on how to use it (2).

By moving through the various elements of the tree, the facility can be analyzed or designed for fire safety. As an example, the "self-termination" objective (Fig. 8.1) can be easily controlled by the designers. Geometry, fuel, and ventilation can be manipulated through the design process. The height and volume of the room; the amount, volume, and distribution of furnishings; the flame spread factor; and the type of wall finishes will have effects on a fire.

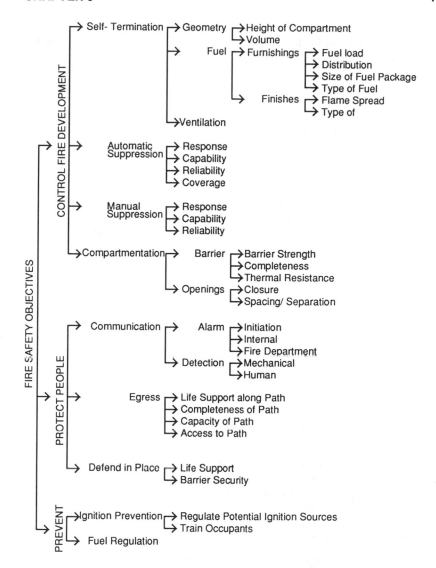

Figure 8.1 Fire protection tree. (Courtesy of Department of Commerce.)

Design decisions have an effect on the facility's fire protection, which creates a compelling argument for the early introduction of fire protection engineering into the design process.

SITE SELECTION

Site features should be considered in the overall fire safety design, as well as in the other design subsystems. The site characteristics can play a major roll in the type of fire defenses used for the facility. The major site features follow

Water Supplies

From the viewpoint of a fire protection engineer, one of the main considerations in plant site selection is the availability and strength of the water supply. Although water supply information is only one of many inputs that must go into site selection, this input can be of utmost importance because costly expenditures may be required if an adequate municipal supply is unavailable. Water supplies will be covered in more detail in Chapter 9, Water Supplies for Fire Loss Control.

Traffic and Transportation

Time plays a major role in the response of the public fire department to the facility. Traffic and the arrangement of the streets in the community have an effect on the response time. Long fire department response times may signal a need for a fully equipped internal fire brigade.

Public Fire Protection

The type of the protection, whether it is paid, part-paid, or volunteer; the location of the fire stations and the anticipated response time; and the nature and condition of the fire-fighting equipment and the alarm system need to be considered. Access to the site is very important. Built-up, congested areas may include other properties, limiting access to the site. In addition, the topography of the site may limit fire department access.

Exposures to the Site

The facility could be exposed to conflagration hazards from a large concentration of combustibles within close proximity, such as hazardous chemical risks, oil refineries, large quantities of piled lumber, etc. During a fire, exposure can occur from horizontal radiation and from flames coming from the roof or top of a lower burning building. A number of factors—such as the intensity of the exposing fire, the duration, the total heat produced, the features of exposed and exposing buildings, the air temperature and humidity, the wind velocity and direction, and fire department access—significantly influences the danger of an exposing fire. A number of detailed guides have been established to quantify the exposure hazard. One of these is the *Recommended Practice for Protection of Buildings from Exterior Fire Exposures*, NFPA 80A (2).

Exposures to the Community

As stated in an earlier chapter, the EPA has established regulations concerning the community's right to know of the processing, storing, manufacturing, and use of hazardous materials by industry. In addition, the community may also have noise control ordinances, explosives ordinances, and other types of environmental regulations. The site should be planned to take into account the total effect on the surrounding community.

Flood Plain

Fire hazards are increased during flooding, and fire department access is severely restricted. Careful attention should be directed to whether the site is in a known flood plain. For instance, being in a 100-year flood zone may seem fairly secure, but, in fact, works out to be a 63% chance of being flooded within the 100-year period.

Information on flood-plain areas can be obtained from the U.S. Geological Survey flood maps, U.S. Army Corps or Engineers reports, Tennessee Valley Authority reports, and Federal Insurance Administration Flood Hazard Boundary Maps and Flood Insurance Studies. Tied in with this general subject is the nature of surface drainage in the area, which could possibly create flash-flooding conditions (5, 6).

Earthquake Zone

A large amount of the total damage caused by earthquakes may be due to the fires and explosions that follow. The site should be assessed for its location in a possible earthquake zone. A number of design modifications would be necessary if the site is located in such a zone. In a recent earthquake in the San Francisco Bay area (October 1989) that measured 7.0 on the Richter scale, the buildings designed to be earthquake resistant had little or no damage, particularly the high-rise buildings while the conventionally designed older buildings had considerable damage.

Fault locations can be identified from fault and seismic risk maps available from the U.S. Geological Survey, *Uniform Building Code*, National Science Foundation, and the American National Standards Institute (7).

PLANT LAYOUT

Fire protection generally requires compromises. Compromise, which may seem inappropriate, is based upon the necessity of being able to operate a plant efficiently incorporating engineering design features that, at the same time, will make the facility as fire safe as practical. For example, fire protection engineers often must be content with having a hazardous operation located in the center of a plant building, where explosion-relief venting can be obtained only through the roof, rather than at the periphery of a building, where explosion relief can be more readily obtained through side walls and the process can be more readily enclosed. Other factors such as process flow may dictate the exact location of the hazardous operation.

While the interior layout of a single-building plant can present some problems, the layout is relatively simple and clear-cut from a fire safety standpoint. Layouts become somewhat more complicated, again from a fire safety standpoint, when one is involved with complex, multibuilding plants involving hazardous processing, such as petrochemical facilities. Here, consideration must be given to the concentration of these hazardous processes; the accessibility; the location and protection of the vital utilities and control rooms; and the location of employee amenity areas, such as offices, locker rooms, lunchrooms, etc.

Important considerations involved in plant layout are as follows:

1.   The separation of hazardous operations or processing by distance or building separation. If explosion hazards are involved, it must be determined that separation by distance is adequate or, as a less desirable alternative, protective measures should be considered, including the construction of substantial explosion-resistant barricades. The adequacy of the distance must be determined by considering the nature of the hazard involved from a fire and explosion standpoint (possibly a vapor cloud analysis) and also by considering the exposure to loss of unusual value concentrations and specific problems involving the possibility of disruption to the continuity of operations; i.e., the criticality of a part or unit of the production chain.

2.   Accessibility for manual fire suppression. This will determine the adequacy of roadways or other means of access by fire equipment to production units. The inability of fire equipment to reach certain locations and provide effective fire fighting will often dictate the need for further automatic protection, such as automatic water-spray systems and monitor nozzles either remotely or manually controlled.

3.   The separation of important and vital plant utilities such as boiler rooms, electrical substations, and control rooms, from exposure to fire or explosion. In multibuilding operations, it is generally considered that the safest location for such facilities will be on the periphery of the plant site.

4. Location of employee amenity areas, as mentioned earlier, such as locker rooms, cafeteria, office areas, etc. From a standpoint of life safety these areas should be located on the periphery of the facility.

5. The extent and concentration of values. While it is generally desirable to group more hazardous operations together, consideration should be given the concentration of values and separation to avoid the possibility of unacceptable losses. In single-building locations, this will usually involve the separation by fire walls of high-valued storage areas from processing areas.

6. Topography. This is an extremely important consideration where plant operations involve the use of extensive volumes of flammable liquids. In these instances, consideration must be given to locating vital production units and utilities so that they will not be exposed by possible flows from the release of flammable liquids under emergency conditions. Where it is impractical to locate these units at elevations not subject to flammable liquid flows, the use of diversionary curbs or walls to direct flammable liquid flows to safe impounding areas or confinement of flammable liquids by diking is advisable.

Large plants, and indeed every facility, require careful consideration of the plant layout, taking into account the elements discussed above. At the same time, one must recognize the problems brought about by modern technology; the increasing tendency to a greater concentration of values and hazards; the need for a high degree of employee training in handling complex processing, and the very important need, particularly in complicated, high-valued chemical plants, for preventive maintenance programs of the highest quality.

FIRE PROTECTION PLANS

The ability to read plans is important in all phases of engineering, and the fire protection field is no exception. It is essential to be able to visualize a facility that one has not seen or, even if the facility has been visited, to be able to show on a sheet of paper the exact construction, protection afforded, location, and size of fire protection mains and connections; the location and type of water control valves; the location and types of fire pumps; the location and thicknesses of fire walls with types of penetration protection; and other vital data.

In the fire protection field, the plans are unique in that one drawing will show all of the above, whereas in other phases of engineering, where more detailed information is required, there will be architectural, mechanical, structural, electrical, and plumbing plans, covering many sheets. This single plan, in addition to using symbols unique to fire protection, is usually colored to show at a glance the construction of various buildings.

Figure 8.2 shows examples of the plan symbols generally in use in fire protection (1). Readers familiar with standard engineering plans will note that the fire protection symbols differ from engineering symbols as typified by the symbols

Figure 8.2 Plan symbols generally used in fire protection. (Reprinted with permission from the *Fire Protection Handbook*, 16th Edition, Copyright 1986, National Fire Protection Association, Quincy, MA 02269.)

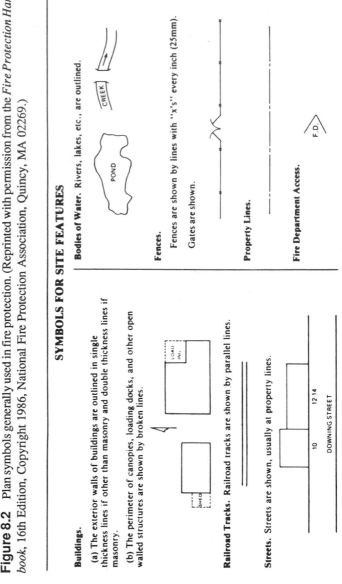

SYMBOLS FOR SITE FEATURES

Buildings.

(a) The exterior walls of buildings are outlined in single thickness lines if other than masonry and double thickness lines if masonry.

(b) The perimeter of canopies, loading docks, and other open walled structures are shown by broken lines.

Railroad Tracks. Railroad tracks are shown by parallel lines.

Streets. Streets are shown, usually at property lines.

Bodies of Water. Rivers, lakes, etc., are outlined.

Fences.

Fences are shown by lines with "x's" every inch (25mm).

Gates are shown.

Property Lines.

Fire Department Access.

SYMBOLS FOR BUILDING CONSTRUCTION

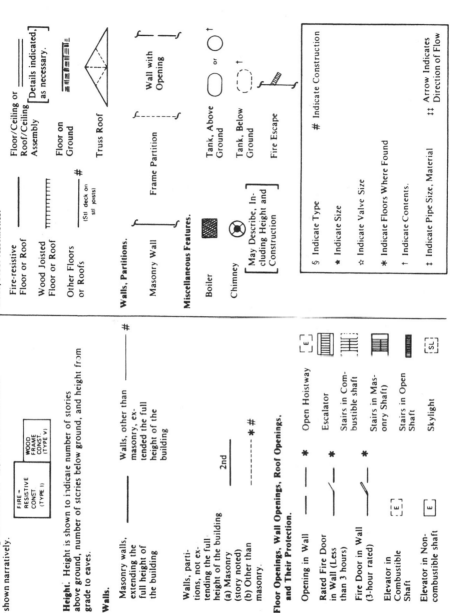

Types of Building Construction. Types of construction are shown narratively.

FIRE-RESISTIVE CONST (TYPE I)	WOOD FRAME CONST. (TYPE V)

Height. Height is shown to indicate number of stories above ground, number of stories below ground, and height from grade to eaves.

Walls.

Walls, extending the full height of the building

Walls, other than masonry, extended the full height of the building ── #

Walls, partitions, not extending the full height of the building

(a) Masonry (story noted) — 2nd

(b) Other than masonry ─ ─ ─ * #

Floor Openings, Wall Openings, Roof Openings, and Their Protection.

Opening in Wall ── ── *

Rated Fire Door in Wall (Less than 3 hours) *

Fire Door in Wall (3-hour rated) *

Elevator in Combustible Shaft [E]

Elevator in Non-combustible shaft [E]

Open Hoistway [E]

Escalator

Stairs in Combustible shaft

Stairs in Masonry Shaft)

Stairs in Open Shaft

Skylight [SL]

Roof, Floor Assemblies.

Fire-resistive Floor or Roof

Wood Joisted Floor or Roof

Other Floors or Roofs ── # (Stl deck on stl joists)

Floor/Ceiling or Roof/Ceiling Assembly [Details indicated, as necessary.]

Floor on Ground

Truss Roof

Walls, Partitions.

Masonry Wall

Frame Partition

Wall with Opening

Miscellaneous Features.

Boiler

Chimney

Tank, Above Ground

Tank, Below Ground

Fire Escape

[May Describe, Including Height and Construction]

§ Indicate Type

* Indicate Size

☆ Indicate Valve Size

* Indicate Floors Where Found

† Indicate Contents.

‡ Indicate Pipe Size, Material

Indicate Construction

‡‡ Arrow Indicates Direction of Flow

SYMBOLS FOR WATER SUPPLY AND DISTRIBUTION

Mains, Pipe.

Public Water Main

Private Water Main

Water Main Under Building

Suction Main

Hydrants.

Private Hydrant, One Hose Outlet

Public Hydrant, Two Hose Outlets

Public Hydrant, Two Hose Outlets and Pumper Connection.

Wall Hydrant, Two Hose Outlets

Private Housed Hydrant, Two Hose Outlets

Indicate Type, Construction, Size and Height via Notations

Indicate Type, Construction and Size via Notations. Symbol Orientation Must Not Be Changed.

Stored Water.

Water Tower or Tank-Above Ground

Pressure Tank

Meter

Valves.

Post Indicator and Valve

Key-operated Valve

Valve In Pit (OS&Y Shown)

Check Valve

Non-indicating Valve (Non-rising-Stem Valve)

OS&Y Valve (Outside Screw and Yoke, Rising Stem)

Fire Department Connections.

Two-way Fire Department Connection (Siamese Connection)

Freestanding Fire Department Connection (Siamese Connection)

(Specify Size and Angle)

(Sidewalk or Pit Type, Specify Size)

Fire Pump.

Fire Pump with Drives

SYMBOLS FOR SPRINKLER SYSTEMS[1]

Piping, Valves, Control Devices.

Sprinkler Riser

Check Valve, General

Alarm Check Valve   or

Dry Pipe Valve   or

Dry Pipe Valve with Quick Opening Device (Accelerator or Exhauster)

Deluge Valve

Alarm/Supervisory Devices.

Flow Detector/Switch (Flow Alarm)

Pressure Detector/Switch   § (Specify Type – Water, Low Air, Hi Air, etc.)

Water Motor Alarm (Water Motor Gong)

(Shield Optional)

Electric Alarm Bell (Electric Alarm Gong)

SYMBOLS FOR EXTINGUISHING SYSTEMS

Wet (Charged) System.

(a) Automatically Actuated   or

(b) Manually Actuated

Dry System.

(a) Automatically Actuated   or

(b) Manually Actuated

Foam System.

(a) Automatically Actuated

(b) Manually Actuated

For Liquid–, Gas–, and Electrical-type Fires.

(b) Automatically Actuated

For Fires of All Types, Except Metals.

(a) Automatically Actuated

(b) Manually Actuated

Carbon Dioxide System.

(a) Automatically Actuated

(b) Manually Actuated

Halon System.

(a) Automatically Actuated

(b) Manually Actuated

Supplementary Symbols

Nonsprinklered Space

Partially Sprinklered Space

[1]These symbols are intended for use in identifying the type of installed system protecting an area within a building.

Figure 8.2 Continued

for valves. The fire protection symbols are quite simple and rather universally understood.

Colors are used more often than not to show construction types, although it is probably reasonable to say that there is a trend away from coloring because of the cost involved. While these colored plans present at a glance the construction types and mixes, this information is also available by a closer review of uncolored fire protection plans where not only the plan view, but also sections show the roof, wall, and floor construction. It is extremely important that a fire protection plan be complete and show all the salient features of a facility; this certainly includes adequate cross-sectional views.

The more generally used colors for various constructions are as follows:

Yellow: Frame (Type V)
Pink: Heavy timber/ordinary (Type III and IV)
Gray: Noncombustible/limited combustible (Type II)
Gray with brown border: Noncombustible/limited combustible with masonry
    walls
Light brown: Fire resistive (Type I)

Due to slight variations in the use of standard symbols, there may be some slight variation in plans. However, these differences are so small as to be of no consequence to the person reviewing the plan.

A relatively simple fire protection plan is shown in Figure 8.3. Of particular interest, other than the structural details, is the arrangement of the underground fire mains and water supply connection with so-called isolation valves located in the underground main system. Many of the features shown on the plan are explained in the various chapters of this book.

Plans of the type under discussion are usually obtained from HPR underwriters, but the following noninsurance sources are also available to furnish diagrams of this nature.

Sanborn Map Co. Inc.   Holmes-Watson Drafting
    629 Fifth Avenue   2721 Berkshire Drive
    Pelham, N.Y.   San Bruno, Calif.

With a basic understanding of the principles of fire prevention and protection and knowing where additional reliable information can be obtained, the person responsible for controlling fire losses can make an immediate evaluation of hazards by carefully studying a fire plan. It is an effective and practical approach to the thorough and detailed analysis that is always advisable.

CONSTRUCTION MATERIALS

The purpose of this section is to indicate to architects and engineers the availability of information pertaining to the fire characteristics of various building materials and components.

Classifications and Materials

It is important to briefly examine the various building classifications as accepted by the fire protection industry and as included in the *Standard for Types of Building Construction*, NFPA 220 (2). The construction types identified in NFPA 220 and in the three model building codes—i.e., *Uniform Building Code*, *Basic/National Building Code*, and *Standard Building Code*,—are derived from five fundamental construction types: fire resistive, noncombustible, ordinary, heavy timber, and wood frame. These descriptive names are being discontinued because they do not define the construction types as precisely as needed in the current design methods and architecture.

The construction types currently in use have Roman numerals for the type and a notational system for the fire resistance rating. The fire resistance rating (in hours) is for three basic elements of construction: exterior wall, structural frame, and floor. For example, Type I (443) would indicate a fire-resistive building with 4-hr exterior walls, 4-hr structural frame, and 3-hr floor construction. Refer to NFPA 220 for fire resistance requirements for the various types of construction (2).

Limited Combustible

NFPA 220 now realistically defines noncombustible building materials as those that "will not ignite, burn, support combustion, or release flammable vapors (2). It is interesting to note that before 1975, the definition of noncombustibility included materials having a structural base of noncombustibility with a surfacing not over 1/8 in. thick with a flame spread not exceeding 50.

The former inclusion of these materials under the definition of noncombustible led to problems. Certain materials, particularly polymeric ones, were included in the category of noncombustibility, although they possessed definite flammability characteristics under fire conditions. To fill a gap existing between true noncombustible materials and those building components possessing a very low degree of combustibility—and, as a consequence, acceptable in many building applications,—the term *limited combustible* came into use. This term now includes those materials that are not noncombustible under the above definition, with NFPA 220 (2) indicating materials having a potential heat value not exceeding 3500 Btu/lb possessing either:

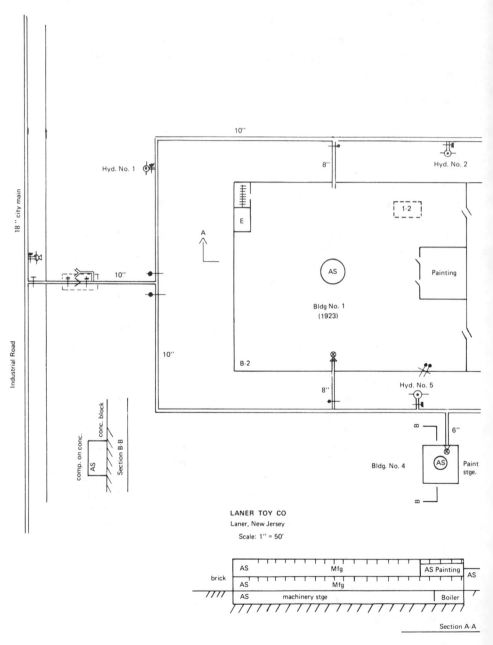

Figure 8.3　A typical fire protection plan of an industrial facility.

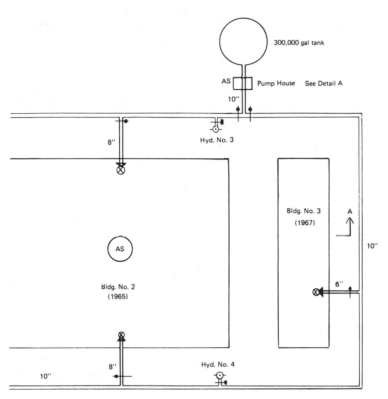

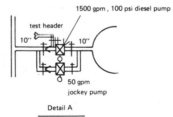

Detail A

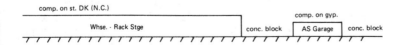

1. A structural base of noncombustible material with a surfacing not exceeding 1/8 in. with a flame spread of 50 or less, or
2. A flame spread of 25 or less not evidencing continued progressive combustion.

The term *potential heat value* is probably unfamiliar, to has a significant importance where polymeric materials are encountered. This term is defined by NFPA 220 (2) as "the average value, in Btu per pound, obtained by testing a building material in accordance with *Tentative Method of Test for Potential Heat of Materials in Building Fires* (8). It should be noted that the higher potential heat value of a material, the greater the flammability of the substance.

Fire Resistance Ratings

Before proceeding further with our discussion of types of construction, it seems advisable to define *fire resistance rating*: the length of time for which a structure or structural members can resist a fire, usually before each can no longer support loads. The fire resistance ratings of masonry units to composite assemblies of structural materials for buildings including bearings and other walls and partitions, columns, girders, beams, slabs and composite slab and beam assemblies for floors and roofs, are obtained from fire tests conducted by nationally recognized testing laboratories in conformance with the *Standard Methods of Fire Tests on Building Construction and Materials* NFPA 251 (2). For more detailed information, refer to this standard for full understanding of the tests and the failure points.

The UL publishes annually, as a result of its testing program, two very useful publications for architects, engineers, and those interested in the field of fire protection, entitled *Fire Resistance Directory* (9) and *Building Materials Directory* (10). The UL *Building Materials Directory* will be discussed later and pertains basically to the flammable characteristics of various building materials. The UL *Fire Resistance Directory* lists fire resistance ratings expressed in hours and applied to floors, roofs, beams, columns, and walls with a complete description of the structural component. The detail test method and criteria used for establishing fire resistance ratings are described in *Standard Methods of Fire Tests of Building Construction and Materials*, NFPA 251 (2) and UL 263 (11). The designs tested are indicated in the publication by a design drawing along with a description of all the components. These designs are identified by four-digit alphabetical-numerical design numbers, with the prefix letter designating the type of protection and the final two numbers identifying the particular design. Various groups of construction are provided with prefix letters in the publication, as listed in Table 8.1.

In using the UL *Fire Resistance Directory* (9), it must be noted that the ratings are applicable only to assemblies in their entirety. In tests, individual components are not assigned fire resistance ratings, but are part of the overall design. In de-

Table 8.1 Groups of Construction

Designated prefix letters	Group of construction
A, B, or C	Floor-ceiling designs: concrete and (cellular only) steel floor units
D, E, and F	Floor-ceiling design: concrete and steel floor units
G, H, and I	Floor-ceiling designs: concrete and steel joist
J or K	Floor-ceiling designs: concrete
L or M	Floor-ceiling designs: wood joist
N or O	Beam designs for floor-ceiling assemblies
Q or R	Roof-ceiling designs
S or T	Beam designs for roof-ceiling assemblies
U, V, or W	Wall and partition designs
X, Y, or Z	Column designs

signs where structural steel components are used, minimum sizes are specified, but sizes having greater thicknesses may be used.

In the UL *Fire Resistance Directory* (9), composite roof and floor assemblies have two means of obtaining fire resistance ratings: one by direct application of the protective material to the steel supports and the other by providing a so-called membrane protection on the underside of the assembly, which will also become the ceiling surface. The membrane method where a ceiling is desired, in all probability, is less costly, with protection taking the form of plaster, gypsum board, or specified acoustical tile ceiling systems using steel framing suspension.

Sprayed-on Protection

Structural steel must be protected (insulated) from exposure to heat in a fire (fireproofing). This can be done by encasement of the steel, application of a surface treatment, or installation of a suspended ceiling (floor-ceiling assembly) capable of providing fire resistance. Due to costs and the lighter weight of the added material, surface treatments are a very popular alternative. At one time, a number of buildings used a sprayed-on material containing asbestos, which was an excellent heat insulator. Unfortunately, asbestos was determined to be a carcinogen, and flaking of this sprayed-on application created a public health threat. The federal government (EPA) came out with strong regulations prohibiting the use of asbestos and required the removal of asbestos sprayed-on material that posed a health threat. Removing the asbestos material and reprotecting the steel members created a major expense. Now, sprayed-on mineral fiber coatings and cementitious coatings are being used. These rely heavily on effective application, complete coverage, and long-term maintenance. Where there is any question, local authorities should be consulted. In any event, surfaces upon which sprayed materials are applied must be free of dirt or scale. Frequent checks of the thicknesses and densities of the sprayed-on protection should be made.

Intumescent paints and coatings have also been used to increase the fire endurance of structural steel. These materials can intumesce (swell) when exposed to flame, transforming into a thick insulating mat.

Type I

Type I (fire resistive) identifies construction where the structural members (walls, columns, beams, floors, and roofs) are noncombustible or limited-combustible, with various fire resistance ratings for the structural members. This classification has two subclasses: Type I (443) and Type I (332). The difference is in the fire resistance ratings.

While there are many forms of Type I construction, this class can be broken down into two main categories: monolithic reinforced concrete construction and protected steel frame construction. Monolithic reinforced concrete construction is defined as a building having no structural steel framing. A structural steel framework is provided with fire resistance by concrete or other materials.

Examples of monolithic fire-resistive construction are flat slab, beam and girder, beam and slab, rib type (whether one-or two-way ribs), and flat plate construction.

An architect's determination of the type of construction to use will generally be influenced by the floor-load capacity and by cost. It can probably be said that flat slab, fire-resistive construction, very popular in the 1920s for multistory, high-floor-load buildings, is infrequently found in modern design. This type of construction features floors and roofs of uniform thickness except at columns, where additional thickness is provided by drop panels and column capitals to transfer the load to the columns and to prevent shearing.

In beam and girder construction, the roof and floor loads are transferred to reinforced concrete beams, from the beams to reinforced concrete girders, and from the girders to columns.

Rib-type, flat plate, and slab-band fire-resistive constructions are designed for lighter floor loads. Flat plate and slab-band types are used particularly in office buildings, schools, etc., where relatively light floor loads would be anticipated.

In the design of reinforced concrete members, whether they are the actual floor or the floor supports in the form of beams and girders, the designer must consider compression and tensile forces, which come into play in these members when they are loaded. Concrete has good compression strength but virtually no tensile strength (tensile strength being defined as the ability to withstand forces that tend to pull the material apart). To provide a structurally safe design, reinforcing steel bars are added to the concrete where tensile stresses will develop. In a simple beam or girder, these tensile stresses will develop in the bottom portions of the center of the structural member and at the top of the structural member where it passes over a column. In these members, the most important consideration from the standpoint of the fire resistance is the protection of the steel reinforcing bars. It should therefore be kept in mind that when concrete spalls and exposes the reinforcing bars, this structural element no longer has the original degree of fire resistance.

Today, many structures of fire-resistive construction are being erected using prestressed, precast concrete. Roof and floor design depicting single and double-T or stemmed prestressed, precast concrete members (and other precast concrete roof and floor assemblies) are shown in the UL *Fire Resistance Directory* (9). The constructions have varying fire resistance rating depending upon the minimum slab thickness and the acceptability of precast concrete units for the particular design. Prestressed concrete units fare available in two general types, known as pretensioned and posttensioned. In pretensioned units, the tendons (reinforcing steel) are prestressed in place before the concrete is poured. In posttensioned, prestressed concrete, tendons are placed on forms before casting, placed in flexible metal tubes, or pushed through holes or cores in concrete by removable cores. The posttensioned type may also be either bonded or unbonded, with the bonded type providing grout that is forced into the space around the tendons after stressing. Prestressed concrete has the advantages of providing smaller deflections and smaller dimensions for the same loading condition, being generally crack-free, and providing smaller loads on supporting members.

Pretensioned and posttensioned structural members show practically equal performance in fire tests. However, the steel used in prestressing is more sensitive to fire and more adversely affected by a given fire exposure than is conventional reinforcement. Usually, cold-drawn high-strength steel is used; if this steel is exposed to temperatures of approximately 750°F (399°C), a large part of the normal strength is lost. Prestressed concrete units use higher quality concrete that is denser and more subject to spalling. Another consideration is that there is usually no continuity of steel reinforcing in prestressed concrete, and prestressed concrete members are often not provided with proper protection of end pieces where cut wire, anchor plates, and bolt heads are located. This is a point of weakness that should be considered for protection in any design.

Type II

Type II (noncombustible) identifies construction where the structural members are noncombustible or limited-combustible, with fire resistance ratings less than those of Type I construction. This classification has three subclasses: Type II (222), Type II (111), and Type II (000). Type II construction will generally afford adequate fire safety for residential, institutional, business, and assembly occupancies (2).

Metal-Deck Roofs   Under this type of building classification, the use of insulated metal roof decks is quite common, primarily because of relatively low cost. This type consists of a roof covering on insulation, held by an adhesive with or without a vapor barrier on a steel roof deck. Insulated metal-deck roofs can be considered either noncombustible or combustible depending upon construction details.

When this roof first came into prominence in the late 1940s and 1950s the roof deck was initially considered to be noncombustible under all conditions. It took several fires including the classic General Motors Livionia transmission plant fire in 1953 to point out the combustibility of the normal roof. At that time, such a roof contained a considerable quantity of combustible asphaltic materials between the insulation board and the roof deck. Under fire conditions, this asphastic material vaporizes unburned gases and forces them down between the joints of the steel deck, since they cannot escape through the insulation board and roof covering. As a consequence, it is possible, with a roof having considerable qualities of these asphaltic materials, to flash a fire across the underside of the roof deck. It can be generally concluded that combustible materials above the insulation board are not a factor in fire losses unless foamed plastic insulation is used.

The components of a metal-deck roof assembly consist of the steel deck, the adhesive or mechanical fasteners, a vapor barrier, if any, and the insulation. These roofs are classified as Class I or Class II depending upon the heat-release rate. Class I is considered to be noncombustible and does not require sprinkler protection for the roof construction itself. Class II is considered to be combustible, again because of the large quantities of asphaltic materials used for adhesion.

To reduce the flammability of a design, one of the following courses of action must be taken:

Replace the combustible asphaltic materials with a noncombustible adhesive

Replace the combustible asphaltic adhesive with mechanical fastening in the form of serrated nails or clips for the entire roof (generally not done except in extremely high buildings or at roof perimeters).

Reduce the quantities of the asphaltic roof materials to the point where heat-release rates will be within acceptable limits. In general, this means limiting the use of the asphaltic materials to 15 lb per square (100 ft^2), although 12 to 25 lb per square is acceptable when applied in ribbons 2 to 3 in. wide.

More information on metal-deck roofs can be obtained from the Factory Mutual *Less Prevention Data Books* (12). UL has also conducted a considerable number of tests on metal-deck assemblies; these assemblies are listed in the UL *Building Materials Directory* (10) as being "fire-acceptable," with others tested for wind uplift only indicating the design in pounds per square foot uplift for the particular roof. The Factory Mutual *Approval Guide* 13, under the "Building Materials and Construction" section, lists approvals for roof deck components.

Approved listings of insulation are given, in the FM *Approval Guide* (13) and UL *Building Materials Directory* (10), consisting of wood fiberboard, perlite board, glass fiberboard, cellular glassboard (foamed glass), and composite board. Shortages in rigid insulation have been quite common in recent times; as a consequence, there has been a noticeable trend toward the use of the more available polyurethane rigid insulation. This material, being extremely flammable, should be avoided unless used in a composite board where it is adhered to and protected by a noncombustible board. Two of these composite boards containing polyurethane, which are acceptable, are listed in the FM *Approval Guide* (13).

Vapor barriers, used in metal-deck roof assemblies provide protection to the insulated metal-deck components above the metal deck from condensation due to moisture within the building. It should be noted that Class I metal-deck roofs do not require vapor barriers, which are generally polyvinyl chloride sheeting.

Designers should consider the roof-deck gauge configuration and span between supports. Smaller gauges and greater spans will result in greater roof deflection, causing the breakdown of the bond between the insulation and roof deck and a greater possibility of windstorm damage. Roof decks are constructed of rib or corrugated configurations to prevent flex and are usually 1.5 to 2 in deep and not lighter than 22 gauge.

The matter of windstorm damage is of considerable concern in insulated metal roof decks, particularly Class I types where quantities of the asphaltic materials are reduced to obtain acceptable burning rates. This, in addition to poor application of the adhesive, leads to above-average windstorm problems unless additional precautions are taken, particularly at the roof perimeters where the majority of windstorm problems originate. Additional protection takes the form of providing mechanical fastening for the roof perimeter using serrated nails or clips such as indicated in the FM *Approval Guide* (13). It is generally considered that providing mechanical fastening at the roof perimeter for a 4-ft depth is adequate. However, extremely high buildings or buildings located in unusual areas subject to above-average winds may require fastening of larger areas. Designs for extremely high insulated metal-deck roofs might well consider mechanical fastening of the entire roof deck.

Noncombustible Formboard   Another popular type of noncombustible construction involves the use of gypsum or lightweight concrete either on metal forms or decks or on formboard supported by unprotected steel framing. The use of combustible formboards has been a contributing factor in some large losses, including a large grocery warehouse fire that occurred in a Boston suburb where a gypsum roof was poured on combustible formboard. Acceptable formboards are listed under the "Minerals and Formboard" category of the UL *Building Materi-*

als Directory (10), although from a fire protection standpoint any noncombustible board can be used.

Type III

Type III (exterior protected combustible or ordinary) identifies construction where the exterior walls are noncombustible or limited-combustible and the interior structural members may be combustible. This class has two subclasses: Type III (211) (protected) and Type III (200) (unprotected) (2).

This type of construction, which requires minimum 2-hr exterior bearing walls or nonbearing walls to be of noncombustible construction, has roof, floors, and interior framing wholly or partially of wood or other combustible material. Type III construction will withstand a fire for only about 15 min. The fire resistance rating can be increased by sheathing the underside of joists with gypsum board or similar material (which at the same times introduced concealed spaces between the joists). Such concealed channels, if existing, should be provided with noncombustible barriers or fire stops. This type of construction also requires fire stopping at interior studding and where joists are tied into the exterior walls. Exterior wall corbeling, which is the laying of the masonry to protrude beyond the interior wall line just below the joist bearing, may be used for fire stopping at floor levels. At the same time, joists supported by masonry walls should be cut on an angle to provide a so-called fire cut, which will allow the joists to self-release under fire conditions without pulling away sections of the wall.

Type IV

Type IV (heavy timber) identifies construction where the structural members are of unprotected wood (solid or laminated) with large cross-sectional areas. Interior and exterior walls are of noncombustible or limited-combustible construction. This type of construction was historically known as heavy timber or "mill" construction, owing its name to the textile factories and paper mills built during the mid-1800s. While not considered economical under today's conditions, one-story masonry or other 2-hr-rated, walled buildings with 2-in.-thick wood plank decks supported by laminated wood arches do fall into this category and are relatively common (2).

Points of weakness in heavy timber construction are sharp edges of timber, such as the edges of columns, beams, and girders; the base of columns; and splices in roof trusses, because fire attacks the edges first. Weak points can be improved by chamfering or rounding corners of the structural members, by protecting the base of columns with noncombustible supports such as cast iron pintles, and by providing minimum 2-in. wood plates under splices to prevent the attack of fire from beneath.

Type V

Type V (wood frame) identifies construction where the structural members are of wood or other combustible material. This type has two subclasses: Type V (111) (1-hr protection throughout) and Type (000) (no fire resistance requirement).

There are two basic types of frame construction, "plank and beam" framing and the conventional "platform" framing. With both, fire stopping is essential in concealed spaces (2).

Interior Finish

The *Life Safety Code*, NFPA 101 (2), provides for certain levels of interior finish according to various occupancy classifications based upon the surface burning characteristics or flame-spread ratings. Interior finishes are classified as follows:

Class A interior finish: flame spread 0–25, smoke developed 0–450
Class B interior finish: flame spread 26–75, smoke developed 0–450
Class C interior finish: flame spread 76–200, smoke developed 0–450

The *Life Safety Code* stipulates the required class of interior finish for certain areas of facilities, such as means of egress, places of assembly, etc., allowing higher classifications or flame spreads where sprinkler protection is provided. Earlier, we briefly discussed fire tests of various building components, but did not include tests of surface-burning characteristics of building materials, which are tested using methods described in *Standard Test Methods for Higher Hazard Classifications of Building Materials*, UL 723 (14), and *Method of Test of Surface Burning Characteristics of Building Materials*, NFPA 255 (2). The surface-burning characteristics, including flame spread, fuel contributed, and smoke developed, are given numerical ratings that compare asbestos cement board having a rating of zero and red oak flooring having a rating of 100. A specimen having a flame spread, smoke-developed, and fuel-contributed rating of 50, therefore, has approximately one-half the flammable characteristics of red oak.

Building Materials

A companion UL publication that is indispensable to the architect or engineer involved in construction and also to those in fire protection is the UL *Building Materials Directory* (10). This publication is divided into two parts. Part 1, the Building Materials List, provides the names of companies qualified to use the UL listing mark on or in connection with products. Companies are listed alphabetically under various categories such as air-conditioning systems, through vents, smoke, and heat. Of greater interest to the users of this publication is Part 2, the Classified Building Materials Index. The listings under Part 2 are numerous. Some of the major items of interest to fire protection engineers are the listing of filter units (whether Class I or Class II, depending upon flammability), the listing of fire dampers and their hourly classifications; and the hazard classifications of acoustical materials, batts and blankets (which cover a rather wide variety of products), fire-retardant coatings, mineral and fiberboards, various plastic materials, fire doors, insulated wall construction, etc.

In addition to the UL publications, the FM Approval Guide (13) published annually also has a section on "Building Materials and Construction" that gives fire ratings of various roof and floor assemblies, walls, and partitions and information on roof-deck constructions as previously indicated, insulated metal roof decks, smoke and heat vents, fire doors, etc.

Fire-retardant coatings were mentioned earlier. Intumescent coatings are used to reduce the flame spread of interior finishing materials, which usually reduces cost considerable because replacement of the interior finish itself is not required. The intumescent coatings must be applied in accordance with the directions to get the desired reduction in flame spread. It should be kept in mind that fire-retardant or intumescent coatings do not alter the fire resistance of an assembly, except in the case of a listed intumescent mastic, which can provide up to 2-hour fire rating on structural steel by converting to a thick pad of thermal insulation when exposed to flame. One undesirable quality of the mastic intumescent material is that, when exposed to weather, this material often will not intumesce under fire conditions.

SPECIAL CONSTRUCTION ITEMS

Certain construction features are important to the fire protection engineer in conjunction with special hazards involving explosion—such as blast-resistant construction and explosion-relief venting, the subdivision of buildings and the installation of fire and barrier walls and the protection of penetrations of these walls, and smoke and heat venting. These items will be briefly discussed in this section.

Building Subdivision

Subdivision of buildings is considered to limit the extent of loss by providing smaller fire areas. Subdivision, in addition to limiting the extent of loss, should also be considered to isolate hazardous processes equipment or material and to separate storage from manufacturing or high-valued materials from lower valued stock. Vital building equipment or services, such as power houses, transformers, and record vaults, should also be separated from the main building area to protect the services and the continuity of operations from any fire loss in the main plant area. This will, at the same time, provide reasonable assurance that a fire will be confined to an area such as a boiler room.

The size of fire areas depends upon several considerations, such as the following:

Occupancy: This factor is of prime importance in determining subdivision areas. Low-hazard operations having a low fire loading may occupy large, undivided areas of modern plants, with the opposite extreme required in confining high-hazardous operations to the smallest practical area. The arrangement of the combustibles for fire loading, overall continuity, or quantity of combustibles; the ease of ignition of the combustibles; the presence or absence of ignition sources; and the fire frequency that may be anticipated are other significant factors that determine the maximum size of fire areas.

Values: Where values are quite high, as indicated above, consideration should be given to limiting the extent of loss.

Construction: Obviously, this should be a consideration in determining fire areas. Combustible construction having an occupancy with a low fire loading still requires some reasonable limitation to fire areas.

Sprinkler protection: Larger fire areas may be considered where sprinklers are provided and where either building construction or occupancy require this protection. Even though sprinkler protection is considered in a particular situation to be fully capable of protecting against the particular occupancy or combustible building construction, some limiting of fire areas should be considered, particularly in high-valued properties, because of the possibility of a protection impairment or the inability of sprinkler to control a fire.

Susceptibility of building or contents to manage, particularly to water and smoke: Where highly susceptible and extremely valuable contents are involved, smaller fire areas are advisable.

Protection of extremely vital production units or service areas from the standpoint of being able to maintain continuity of operations and avoid costly shutdowns.

Fire Loading

Fire severity and fire loading are the basis for many of the fire resistance requirements of building codes and government agencies. Even though the fire-loading results tend to be on the conservative side and perhaps technically obsolete, fire loading can be considered a way to determine the degree of fire resistance required for walls or other structural members designed to confine a hazard.

Fire loading may be defined as a measurement of the combustibles within an area. Where only Class A materials (wood, paper, and similar materials) are involved, fire-loading measurements consist of obtaining the total weight of these materials and dividing by the area to obtain an average in pounds per square foot (psf). More accurately this should be done by calculating the caloric content of the combustibles present. Most Class A materials have heats of combustion from

Table 8.2   Heat Values of Some Commonly Used Substances

Substance	Heat of combustion (Btu/lb)
Kraft paper	7500
Corrugated cartons	7700
Hardwood	8700
Softwood	9100
Polyvinyl chloride	8030
Polyethylene	19,950
Polystyrene	18,700
Gasoline (various grades)	19,800–20,520
Acetone	13,228
Kerosene	19,800
Methyl alcohol	9600
Methyl ethyl ketone	14,537
Toluol	18,252

7000 to 8000 Btu/lb, while flammable liquids have approximately twice those values. The heat values for some common substances are given in Table 8.2.

Based on research at the National Bureau of Standards (now the National Institute of Standards and Technology), an approximate relationship can be made between the fire loading and the exposure to a fire, equivalent to the standard time-temperature curve (Fig. 8.4). For example, a 10-lb/ft^2 fire loading in an office building has an assumed release of 80,000 Btu/ft^2 (10 lb/ft^2 x 8000 Btu/lb), which would be the equivalent of a 1-hr exposure under the time-temperature curve. As the fire loading increases, so does the equivalent fire exposure. The fire loading can then be compared to the fire resistance rating of the building. Fire-loading estimates can be found in various references, such as the *Fire Protection Handbook* (1).

Over the years, considerable research in fire loading has been conducted. Some studies have looked at temperature curves of fire severity as it relates to loading, with the ratings (1):

A—Slight fire severity (e.g., offices in noncombustible buildings with noncombustible furniture)

B—Moderate fire severity (e.g., papermaking, institutional buildings)

C—Moderately severe fire severity (e.g., combustible storage areas)

D—Severe fire severity (e.g., manufacturing areas with combustible products)
E—Severe-standard time-temperature fire severity (e.g., flammable liquids, woodworking areas, offices with combustible furniture in combustible buildings)

The British have developed a system using three classifications: low, moderate, and high. Tests have shown that the fire resistance of enclosures can be related to the caloric content using this information as a basis of design of a structure to withstand the fire severity expected, excluding any effect from the discharge of sprinklers. The following fire resistance ratings of enclosures are suggested for the fire severity classes:

Low fire loading: Not exceeding 100,000 Btu/ft^2 for net floor area or 200,000 Btu/ft^2 for limited, isolated areas. This classification will generally apply to schools, offices, hotels, and most areas of hospitals.
Moderate fire loading: 100,000 to less than 200,000 Btu/ft^2 for net floor area or 400,000 Btu/ft^2 for limited, isolated areas. This category will encompass most industrial occupancies, such as machine shops and assembly areas.
High fire loading: 200,000 Btu/ft^2 to less than 400,000 Btu/ft^2 for net floor area or 800,000 Btu/ft^2 for limited, isolated areas. Generally, this classification will cover warehousing of various products.

The British classifications can be compared to the previously mentioned fire severity classes as follows: low—A; moderate—B, C, D; and high—E.

Fire Walls

Fire walls, as opposed to fire partitions, are self-standing walls, having a varying degree of fire resistance, depending upon the hazards involved; having no structural steel framed into the wall; having preferably no wall penetrations, particularly at the upper portion of the wall, and, in most cases, having a fire resistance rating of 3 or 4 hr. These walls are normally of masonry construction (brick, reinforced concrete, or concrete block); however, materials other than heavy masonry may be used. Fire walls should be ground supported and should extend through all stories. Where walls are supported by structural members, special design considerations are required for stresses that may be developed.

Where the roof deck is combustible or where there is a possibility of transmission of heat through the roof decks, such as in an insulated metal deck or an all-metal building, parapets are also required. Parapets should be 36 in. high, but it recognized that on occasion some compromise is required, perhaps for aesthetic reasons. Parapets are not considered necessary above a noncombustible roof deck such as concrete or gypsum.

Openings in exterior masonry walls, on either side of a fire wall, should be protected or blocked up a minimum of 3 ft on each side of the fire wall. Where combustible exterior walls are encountered, the fire walls should end in a T, extending a minimum of 3 ft from each side of the fire wall.

Fire partitions, in contrast to fire walls, generally enclose small areas of hazardous occupancies with the resistance ratings of generally 1 or 2 hr. While fire partitions can be of various materials, for cost reasons, fire resistance in many cases is obtained with the use of layers of gypsumboard either with metal or wooden studding. Metal studding is preferable, so that fire partitions are entirely of noncombustible construction. The fire resistance rating of the fir partitions should be sufficient for the combustible loading, with 1 hr usually required, but with greater ratings normally needed where ordinary combustible loading exceeds 20 lb/ft^2 or flammable liquids exceed 10 lb/ft^2.

Openings in Fire Walls

Openings in fire walls and vertical enclosures such as stair and elevator shafts are the weakest point of any wall or enclosure, so it is desirable when designing a fire wall to have as few penetrations as possible: Fire doors can be designated by an hourly protection rating (ranging from 3 to 1/3 hr), by an alphabetical letter designating the opening (A through E), or by a combination of the two. Openings are classified in the *Standard for Fire Doors and Windows*, NFPA 80 (2), as follows:

Class A: Openings in fire walls and walls that divide buildings into fire areas.
Class B: Openings or penetrations of enclosures of vertical communications, such as stairways and elevators, and in 2-hr-rated partitions.
Class C: Openings in corridor and room partitions, with 1-hr rating or less.
Class D: Openings in exterior walls, subject to severe outside fire exposures.
Class E: Openings in exterior walls, subject to moderate or light outside fire exposure.

Fire doors in interior walls are quite often referred to as Class A, B, or C doors. While it is desirable to have a blank fire wall, this is not practical, but it should be kept in mind that once a wall is pierced, this becomes a weak point and also a maintenance problem.

Fire doors come in various types. The most common variety, in use over a long period of years, is the so-called metal-clad (tin-clad) type, which is constructed of two or three wooden cores covered with 30-gauge steel sheet or terneplate. The metal covering over the wooden cores also protects the nail heads. The two-ply type available in Class B or C ratings carries a 1.5-hour fire resistance rating, with the three-ply type possessing a 3-hr. rating. These doors should have a 3-in.-diameter hole cut in the metal covering (not into the wooden core) on the unexposed side of the door at the time of installation to relieve products of combustion gases that otherwise might cause bulging and rupturing of the door under fir conditions. Unfortunately, this feature is omitted in field installations more often than not. The determination of the "exposed" side is often a matter of judgement; the unexposed side is that side of the door exposed by the lesser amount of combustibles.

Tin-clad doors are generally installed on an inclined track, which is the most simple, reliable, and easily maintained method. Doors so installed close by gravity upon melting of the fusible link, which generally releases a counterweight provided to hold the door open under normal conditions. These doors are also occasionally installed on a level track where an operating weight and opposing counterweight are provided. The melting of a link in this situation releases the counterweight, closing the door. This arrangement is more complicated and less reliable than inclined track installation.

Another type of fire door frequently found in industry is the rolling steel type, which is constructed of interlocking 22-gauge or heavier steel slats. These doors are available for Class A, B, C, D, and E openings. Doors of this type are equipped with coil torsion springs that provide the counterbalancing and closing forces. Two fusible link installations will be found on these doors, one for closing the door under fire conditions and the other for releasing a metal frame stop to prevent the passing of flame over the door.

Composite-type doors, having a noncombustible or fire-resistive core covered with wood or plastic veneer or metal facing, are also in extensive use and are available for Class A, B, C, D, and E openings.

Metal-clad (kalemein) doors were more common a number of years ago and have a metal-covered wood frame of flush or panel design covered with 24-gauge or lighter steel. These doors are available for protection of Class B, C, D, and E openings.

Hollow metal doors are often encountered in modern construction and consist of a metal shell, (20-gauge steel or heavier), filled with noncombustible insulating material. These doors are available for Class A, B, C, D and E openings.

Many of the preceding doors are installed in a swinging or hinged position. Doors for elevators are usually installed in a counterbalance fashion, where they part at midheight or are of the telescoping vertical sliding type.

The selection of fire doors depends upon the use and the clearances provided. Where limited clearances are encountered, rolling steel or swinging types should be used. It is also preferable to have single doors that close the entire opening rather than doors that meet at a center joint and then require an astragal. Using two doors to protect the opening should be considered less reliable.

As indicated above, double (a door on either side of the wall) 3-hr doors should be provided at openings in major fire walls having a fire rating of more than 3 hr. Note that rolling steel doors provide very little resistance to transmission of heat; these doors, therefore, could be undesirable for installation in a fire wall where very high heat release may be expected and where combustibles might be in close proximity on either side of the wall. Where transmission of heat and possible ignition of combustibles on the other side of the wall could be a factor, fire doors bearing the rating "3 hours (A) temperature rise—30 minutes 250°F maximum" should be installed. In other words, the testing of this fire door did not develop a temperature on the unexposed side of the fire door in excess of 250°F (121°C) over a duration of 30 min.

Vision panels, although not permitted, are sometimes found in 3-hr doors, and greatly lessen the resistance of these doors. While not allowed in 3-hr doors, they may be provided in doors of 1.5-hr rating for Class B openings if they are made of 1/4-in.-thick wired glass and the glass area does not exceed 100 in.^{2}. Note that vision panels are also not permitted in doors for Class D openings, preventing severe exposure. Doors installed in partition openings or so-called Class C openings may have 1/4-in. wired glass not exceeding 1296 in.^{2} with no dimension exceeding 54 in.

Door frames and door hardware are important considerations in the installation of fire doors. Doors can be self-closing or can close automatically. Automatically closing swinging and sliding doors have a door holder/release device that releases the door upon activation or an automatic fire detector or a series of pulleys and counterbalanced weights released upon activation of a detector or fusible link.

Some important points in the installation and maintenance of openings in fire walls or partitions are as follows:

Fusible links should be installed properly in the opening where they will be quickly affected by heat. Fusible links should also be protected from mechanical injury.

Where flammable liquids are encountered, raised wills or curbs at fire doors should be provided.

Sills under fire doors should be noncombustible.

Obstructions in fire-door openings that will affect the closing of the door should not be permitted.

Stock or other items should not be stored against, or in the vicinity of, fire doors. Often, this condition prevents the closing of the doors under fire conditions. Where this is a possibility, posts or guards should be installed as a preventive measure.

Fire doors will fail more rapidly if, when closed, stock is stored immediately adjacent to the door. The doors are not designed to provide the listed protection under these conditions; consequently, combustible materials should not be stored near fire doors.

Doors should be checked weekly for proper protection. With doors held open by
weights, this can be done by raising the weight or disconnecting fusible links.
Guides and bearings should be cleaned and lubricated.
Doors should be kept painted. Extreme care should be exercised in painting roll-
ing steel doors to prevent obstructing the door.
Wherever possible, doors should be closed nightly and on weekends.

For a full discussion of fire doors and window, refer to NFPA Standard No. 80
(2) on fire doors and windows and to the section entitled "Fire Doors" in the UL
Building Materials Directory (10).

Construction Considerations

Smoke and Heat Venting

Over the last 20 years, the subject of automatic venting for sprinklered buildings
has generated considerable differences of opinion, particularly among HPR underwriters and fire research experts. Advocates of venting point to the fact that
heat and smoke vents remove the products of combustion, resulting in less water
and cooling being required to absorb the total heat released. This aids in manual
fire fighting and delays the loss of visibility from smoke. Advocates cite the IIT
Research Institute research work as supporting this position, with venting providing two major contributions (1, 15):

Should sprinklers not function properly, the vents would perform as in un-
sprinklered property.
The vents serve as a substitute for manual roof venting by the fire department.

Those opposing vent installation also present a strong position. They point out
that the automatic operation of smoke and heat vents introduces oxygen into an
oxygen-starved environment, possibly causing sprinkler control to be lost due to
an increase in fire intensity and resulting in the opening of a greater number of
sprinkler heads. For instance, the Factory Mutual Research Corporation concluded that these vents are "generally not cost effective in a sprinklered building.
Manually operated vents or other smoke control measures can help after fire control is assured." (1, 16, 17).

Manual smoke and heat venting should be considered in high-piled storage
areas, after control of fire by sprinklers has been established, as an aid to manual
suppression. Automatic venting is highly effective in high-energy-release fires,
such as in flammable liquids, and where large quantities of smoke are given off
under fire conditions encountered in tire and plastic warehouses.

If the major cause of fire spread across the floor of a building is the heat radiat-
ing down from the layer of hot gases beneath the ceiling, venting will limit the

spread of these hot gases. However, if the major cause of fire spread is the flame progressing sideways at the floor level through combustible materials, venting will not prevent this, but will only help to slow it (18).

Smoke and heat venting is usually automatic, actuated by fusible links or may be manual. Manual vents are operated from the roof and permit the release of smoke and heat and the use of hose streams from the roof.

Venting is more effective with the use of draft curtains or draft boards, because the volume of hot gases discharged by thermal updrafts through an opening varies directly with the height of the hot gas columns. Draft curtains may be of any noncombustible material of sufficient strength to be rigid. Corrugated metal is an excellent material for this use.

To establish the amount of smoke and heat venting, it is necessary to classify the type of fire that might be expected, as outlined in the *Guide for Smoke and Heat Venting*, NFPA 204 M (2). These classes are limited to growth fires and continuous-growth fires. Limited growth fires are those that are not expected to grow beyond a predictable size. Continuous growth fires are those that can be expected to grow indefinitely until intervention by fire fighters.

The most desirable spacing of smoke and heat venting would result in a vent immediately over any possible fire. Since this obviously is impractical, spacings have been devised depending upon the type of fire. Recommended vent areas can be calculated by using tables and graphs contained in NFPA 204 M, (2). Smoke and heat venting can be obtained by providing gravity-type vents actuated by the rise of heated air, or by mechanical means using exhaust fans located in the roof.

The most common smoke and heat vent for roofs is the spring-loaded type that is fusible link actuated. These vents should be of an approved type. They are designed to operate against a minimum 5-lb/ft^2 uniformly distributed live load, with snow loading for vents operating against a 10-lb/ft^2 live load. Unit-type vents are available in minimum 48-$\times$ 48-in. size.

Linkage problems in vents of the spring-loaded type have been known to cause failure of vents under fire conditions. As a consequence, NFPA 204 M (2) outlines an inspection and maintenance schedule that includes an acceptance performance test immediately after installation. In addition, a regular schedule for operating the units is also suggested under summer and winter conditions, with a visual inspection of the units not operated during these periods.

Mechanical roof venting often is obtained by using roof exhaust fans installed for employee comfort. It is desirable to mount motors driving the exhaust fans out of the airways to prevent the action of heat directly on the motors. Exhaust fans may be actuated by water flow from a sprinkler systems, by fusible links, or by manual means.

Consideration of the reliability of the mechanical exhaust under fire conditions should include the method of wiring.

The most reliable roof venting is continuous-type gravity vents such as found in the roofs of foundry buildings, which generally are without dampers. These openings are usually unobstructed and have the advantage of no mechanical parts. Roof monitors may also provide automatic smoke and heat venting by using panels with fusible link releases.

Melt-out-type plastic smoke and heat vents use plastic sheeting provided in dome shapes and fabricated to any dimension. Sheeting is stretched over the dome; under heat conditions, the sheeting shrinks and falls out, permitting smoke and heat venting. Steel ribs are usually provided in the dome to catch sheeting and keep it from wrapping around sprinkler piping.

It is feasible to provide mechanical smoke and heat venting by using air-conditioning or ventilating systems in areas such as data-processing installations and in high-rise buildings. This can be accomplished by closing a damper in the main return duct and opening a damper to exhaust under fire conditions. To do this, make-up air must be provided; this is done by opening a damper to permit air intake into the system. For mechanical exhausting, the ventilating system can be actuated either automatically by a smoke-or heat-detection system or manually.

Explosion Venting

Hazards possessing explosion potential necessitate certain structural considerations to confine and reduce damage from explosions. These considerations consist of construction that is resistant to the pressures developed by explosions and provisions for relieving explosive pressures.

Explosion resistant construction may be desirable to protect a vital plant area, a vital plant utility, or an exposure from the possible destructive force of an explosion. Because some industrial processes develop high pressures during explosions, it is economically impractical to design a wall to withstand the full peak force ultimately developed by the explosion (on the order of 100 psig). A normal building wall cannot withstand a sustained internal pressure as small as 1 psig. Therefore, it is advisable to provide explosion-resistant construction along with explosion-relieving construction to permit a lighter design.

In locating a hazardous operation subject to possible explosion, careful consideration must be given so that the operation will not unduly expose personnel or important segments of the plant. Damage can be minimized by locating hazardous operations in outside locations, in small detached buildings. An alternative location is a one-story section of the plant with interfacing walls of explosion-resistant construction.

This addition, preferably should be located at a corner. An extension housing such a hazard should not be constructed adjoining a multiple-story area where it is more difficult to protect the upper portions of the building from the effects of a possible explosion. Interior locations frequently must be used, but these locations should be at an exterior wall where explosion venting can be readily obtained (with the area again separated from the main part of the plant by explosion-resistant construction). Least desirable is an interior location where explosion venting must be completely through the roof. A process in such a location should be enclosed in explosion-resistant construction.

In multistory buildings, a top story location may be used for explosion-producing operations, particularly where dust is generated. Again, such operations should be located at an exterior wall and adequate explosion-relief venting should be provided through the roof and/or the exterior walls.

Explosion-relief venting may be obtained in a variety of ways, all designed to relieve, forces at minimum and maximum pressures of 20 to 25 lb/ft^2. The most desirable type of construction is damage limiting construction. This is a light steel frame metal or other sheathing clad construction, such as aluminum that is self relieving under the force of an explosion.

Where damage-limiting construction is not provided throughout, explosion-relief venting may be obtained through the use of wall or roof members of light construction, generally of lightweight corrugated panels held onto the building framing with special fasteners designed to release at specified pressures. These fasteners are sheet-metal screws with collapsing washers, machine bolts with shank cross sections reduced, or clips.

In addition to the type of wall construction just described, single-thickness glass windows and top-pivoted windows designed to release under explosion pressures may be used. Explosion-venting windows include (1) the top-pivoted type, where the entire window swings out when an explosion occurs and then drops back into position, and (2) the projected type, which has a section that opens when relatively slight pressure forces friction shoes at the top corners to slide down a vertical track while a guard arm permits the section to swing outward to a horizontal position. In determining the amount of explosion venting required to limit pressures resulting from an explosion of a dust, vapor, or gas to prevent major structural damage, certain considerations come into play. One of the most important is the rate of pressure rise of a particular material. With a high rate of rise, there is only a short time for venting the pressure of forces developed by an explosion. A substance having a slow rate of pressure rise can be more easily vented in the event of an explosion. Other considerations are the structural strength of the enclosure; the location of the operation subject to explosion; the type and resistance of the relieving vent itself; and, to some extent, particularly in large areas, the location of the relieving vent.

Larger vent ratios are needed for materials that have high rates of pressure rise. Even larger venting ratios are needed for small enclosures where, because of the size of the enclosure, the entire volume may be involved. In addition, weak structural enclosures require higher venting ratios than do those that are more resistant to damage. In determining the actual venting ratio, it should be kept in mind that,

Table 8.3 Fuel Constant

Fuel	C
Propane, natural gas, gasoline, benzene, acetone, and gases with the same flame speed	2.6
Organic dusts	2.6
Organic mists	2.6
Ethylene	4.0
High flame speed metal dusts	4.0
Hydrogen	6.4

with severe hazards, the highest possible venting ratio based upon the square feet of venting area for the cubic feet of volume should be provided.

Even though the fundamentals of an explosion inside a building can be complex, the Runes equation can be used as a simple method for calculating building venting (19):

$$A_v = \frac{CL_1L_2}{P^{0.5}}$$

where

A_v = vent area (ft^2)
C = constant for fuel (Table 8.3)
L_1 = smallest room dimension (ft)
L_2 = next smallest room dimension (ft)
P = maximum internal pressure that can be withstood by weakest structural member (psi)

and with the constraint:

L_3 less than or equal to 3 times $(L_1 L_2)^{0.5}$
where L_3 = longest room dimension (ft)

The equation holds well for large values of A and low values of P (less than or equal to 3 psi). When using the equation, some odd-shaped buildings (such as L-shaped) may need to be considered as several separate parts.

An example calculation follows for the building with the dimensions in Figure 8.5:

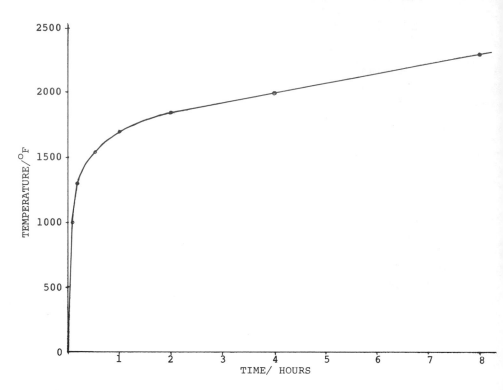

Figure 8.4 The standard time-temperature curve.

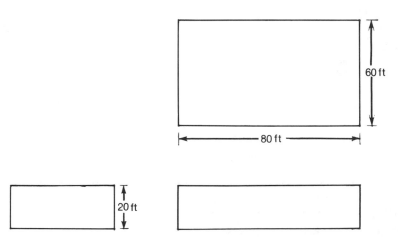

Figure 8.5 Building example for sample venting calculation.

$L_1 = 20$ ft
$L_2 = 60$ ft
$L_3 = 80$ ft
$3 \times (L_1L_2)^{0.5} = 3 \times (20 \times 60)^{0.5} = 103.9$ ft

Since $L_3 = 80$ ft, which is less than 103.9 ft, the space does not have to be subdivided. Assume that the maximum internal pressure, P, is 0.5 psi, and the fuel is ethylene, $C = 4.0$ (from Table 8.3) (19). The vent area is calculated:

$$A_v = \frac{CL_1L_2}{P^{0.5}} = \frac{4 \times 20 \times 60}{(0.5)^{0.5}} = 6788 \text{ ft}^2$$

Since the total available surface = 10,400 sq ft (walls and roof), the venting will take up a major portion of the wall space and some of the roof.

For venting of explosions within vessels or equipment, the "cubic law" that relies on combustion venting nomographs should be used. A detailed explanation of this law, plus sample calculations, can be found in the *Guide for Explosion Venting*, NFPA 68 (2).

In larger volume areas housing hazardous processes subject to explosion, the location of the venting area is important and venting, should be provided as close as possible to the potential release of explosive pressures. It appears that there is little advantage to providing one large venting area as opposed to several smaller vents making up the same total venting area.

Wherever possible, open or unobstructed vents, depending upon climatic conditions, should be provided in lieu of closed or sealed vents, which could be slowed or prevented from opening by ice, snow, or corrosion. Open vents obviously offer no resistance to the relieving pressure.

Sometimes, it is necessary to provide ducts for venting of explosions to the outside such as venting or interior dust collectors. In these installations, the ducts should discharge in the most direct manner to the outside and should also have a high enough structural strength to withstand the maximum pressures that may be developed by the explosion.

PLANT UTILITIES AND SERVICES

Under this topic, we will discuss the fire protection of plant utilities and services, including electrical installations and heating, ventilating, and air-conditioning systems.

Electrical Installations

For general electrical installations, it is advisable for the fire protection engineer to specify that installations be in accordance with the National Electrical Code, NFPA 70 (2), and 29CFR1910, Subpart S, Electrical (20). The code provides for the practical safeguarding of persons and building and their contents from hazards arising from the use of electricity for light, power, heat, radio, signaling, and other purposes. Specifically, the code covers electrical conductors and equipment installed in buildings and other premises and electrical installations adjacent to premises, with certain exceptions such as marine vessels, installations underground in mines, etc.

Certain electrical installations should be discussed as related to fire protection, including electrical installations in hazardous locations, the somewhat closely allied topic of grouping of cables such as may be found in master control rooms, and the safe installation of transformers and switch gear. Electrical maintenance considerations are covered under a later section.

Article 500 of the National Electrical Code (2) defines hazardous locations with respect to the presence of flammable vapors and gases or combustible dust as follows:

Class I, Group A: Atmospheres containing acetylene
Class I, Group B: Atmospheres including gases such as butadiene, ethylene oxide, hydrogen, and propylene oxide
Class I, Group C: Atmospheres including acetaldehyde, cyclopropane, diethyl ether, ethylene, isoprene, or unsymmetrical dimethyl hydrazine
Class I, Group D: Atmospheres including acetone, acrylonitrile, ammonia, benzene, butane, alcohols, methyl ketone, gasoline, and generally other flammable liquids not included in Groups A, B, or C
Class II, Group E: Atmospheres containing metal dust, including aluminum, magnesium, and their commercial alloys and other metals of similarly hazardous characteristics
Class II, Group F: Atmospheres containing carbon black, coal, or coke dust
Class II, Group G: Atmospheres containing flour, starch, or grain dust
Class III: Atmospheres containing easily ignitable fibers or flying but in which fibers or flying are not likely to be in suspension in air in quantities sufficient to produce ignitable mixtures

For a more detailed listing of flammable vapors, gases, and combustible dusts, refer to the *Manual for Classification of Gases, Vapors and Dusts for Electrical Equipment in Hazardous (Classified) Locations*, NFPA 497M (2).

In addition to the preceding classifications, locations are also categorized as Division 1 or Division 2. Division 1 locations are those areas where flammable vapors, gases, or dusts might normally be present in air. Division 2 locations are those areas where flammable vapors, gases, or dusts might be exposed only under upset conditions. For example, a room for storing and dispensing flammable liquids such as acetone, methyl ethyl ketone, etc., or directly transferring liquids from a tank to containers where vapors would be exposed to air would be considered a Class I, Group D, Division 1 location. Conversely, a flammable liquids storage room or flammable liquids processing area where these same materials are handled, but where the flammable vapors are not normally exposed to air except under upset conditions, such as leakage from a drum or pipe breakage, is considered a Division 2 location. It is important to understand and properly classify a location as Division 1 or Division 2 to avoid unnecessary costs in electrical installations.

Explosion-proof electrical equipment is enclosed in a container or housing capable of withstanding an explosion of a specified gas or vapor that may occur within it and of preventing the ignition of a specified gas or vapor surrounding the enclosure by sparks, flashes, or explosion of the gas or vapor within, and that also operates at an external temperature that will not ignite a surrounding flammable atmosphere. Theoretically then, it can be assumed that many explosions take place within an explosion-proof enclosure, but are dampened and prevented from escaping to ignite any flammable vapor dust in the surrounding area. Because of this feature—which is at the very heart of the integrity of the explosion-proof enclosure, openings or other alterations must not be made in the enclosure, and holding bolts of enclosing cases and threaded parts must be screwed tightly.

Underwriters Laboratories publication *Hazardous Location Equipment Directory* (21) contains comprehensive listings of approved equipment for the various hazardous class locations. The Factory Mutual *Approval Guide* (13) also lists equipment of a more specialized nature, for Class I and II locations.

In the various listings, the term *intrinsically safe* electrical equipment will be found; this equipment is approved for hazardous locations. This term refers to equipment or circuits with energy levels, under either normal or abnormal conditions, that are low enough not to cause ignition of the specified explosive atmosphere when the equipment is properly used and installed. This equipment is available for Class I locations only and is covered under a separate heading in the FM *Approval Guide* (13), but is intermixed under various equipment headings in the UL *Hazardous Location Equipment List* (21), such as "Process Control Equipment"

Transformers

From a fire protection viewpoint, transformers present different degrees of hazard and require thought in their installation and location. Aside from the hazard of the transformer itself, this consideration includes the fire exposure that the transformer represents to surrounding areas and the importance of properly protecting transformers to maintain vital power supplies. This second concern frequently is of utmost importance because a loss of a large power transformer can result in a costly business interruption.

The difference in the fire hazard referred to above depends upon the type of transformer. Large-capacity transformers receive high-voltage power from electrical utilities that must be lowered for use within the facility. This lowering of voltage is accomplished within the transformer. Transformers that receive higher voltages and decrease the voltage are called *step-down* types. Transformers that receive lower voltage and increase the voltage are called *step-up* types.

Transformers of the gas-filled, dry type or the high-fire-point, liquid-insulated type—with primary voltage less than or equal to 35,000 volts (V)—are commonly used and installed indoors, since no transformer vault is required.

The gas-filled, dry-type transformers have the conventional dry-type core and coils sealed in a heavy steel tank and filled with fluorocarbon gas for dielectric strength and to provide cooling. Since this type is hermetically sealed, it is safe indoors and outdoors; there is no way for flames or gases to escape if there is a short circuit or failure. This type is available in 1500 to 5000 kilovolt amperes (kVA), 34.5 kV; is recommended for explosive or highly combustible areas; and offers the highest resistance to fire or explosion (22).

High-fire-point, liquid-insulated transformers are insulated with a non-propagating (flame does not spread from source of ignition) liquid having a fire point of not less than 508°F (264°C). These transformers can be used indoors without a vault if not rated at over 35,000 V.

The conventional dry-type transformer does not use a liquid dielectric, but has insulation wrappings of various types. From a fire standpoint, dry types are not comparable in hazard to the oil-insulated type, but some consideration should be given to installations of larger transformers of this type when they are located indoors adjacent to combustibles. While not of any great significance, there is a difference in hazard of dry-type transformers depending upon the degree of combustibility of the insulation used in the windings.

Dry-type transformers located indoors with ratings of 112.5 kVA or less can represent a hazard to adjacent combustibles. To prevent a fire, separation from combustible materials is necessary unless a noncombustible, heat-insulated barrier is provided, or unless the rating of the transformer is less than 600 V with complete enclosure except for ventilating openings. Dry-type units with ratings in excess of 112.5 kVA are required to be installed in a transformer vault or room of fire—resistive construction.

Dry-type transformers rated at more than 35,000 V are required to be located in a vault. HPR underwriters may require enclosures for dry-type transformers of a lesser voltage rating of about 15,000 V.

Dry-type transformers that are designed for outdoor installation require no special treatment other than the usual considerations of safety to personnel and safety of the transformer from an exposure.

Askarel-type insulated transformers use a liquid dielectric, which, until recently, has been polychlorinated biphenyl (PCB). While not presenting a fire hazard, PCBs have been included in the Toxic Substance Control Act., and their use in transformers is required to be phased out over a certain time period in accordance with EPA regulations. This has considerable significance, because askarel-type insulated transformers do not require an enclosure when located indoors or outdoors except when exceeding a voltage rating of 35,000 V.

Fire in oil-insulated transformers occurs from a variety of causes, including a breakdown of insulation, power surges, or failure of a bushing. This type of fire results in a heating of the mineral oil, causing it to vaporize and decompose, with pressure buildup sufficient to rupture the container.

Due to the hazard of the mineral oil in the oil-insulated-type transformers, units of this type should be located in fire-resistive vaults of 3-hr rating with provisions for drainage and for preventing the flow of any released oil from escaping the enclosure. In addition to this consideration for indoor-installed oil-insulated transformers, good practices must be followed in the installation of outdoor power-type insulated transformers exposing buildings or other important facilities. Outdoor units should not be located adjacent to a structure unless the wall is of masonry construction without openings or a minimum 20-ft clearance is provided or the transformer is provided with protection in the form of a water deluge system. Where more than one transformer of large capacity and vital to continuity of operations are closely located, consideration should be given to separating adjacent transformers by fire walls and providing water-spray protection for the larger, more important units. Further, in view of the possibility of the escape of oil, transformers should be located in structures to confine any escaping oil. Dikes or curbs are also important in preventing this situation.

Switch Gear

The installation of switch gear is important in providing for fire safety and continuity of operations. Switch-gear denotes the part of the electrical distribution system that controls power and light and is the main point from which electrical power is distributed and controlled. Modern, metal-enclosed equipment presents very little fire hazard, but from the standpoint of continuity of operations, it should be separated from combustibles either by space or by a fire-resistant, 1-or-hr fire-rated enclosure depending upon the exposure in the form of fuel loading and protection or lack of protection of the exposing occupancy.

Switch gear, whether open or metal enclosed, contains units composed of circuit breakers, motor controllers, and disconnect switches and cutoffs. This equipment is not intended for hazardous locations and, as such, should not be located where exposed to hazardous flammable liquid or dust conditions.

Switch gear for large installations should have circuit breakers separated by sheet metal partitions to prevent excessive damage from an electrical failure. Usually, air type circuit breakers are used, and these present little hazard. Oil-type circuit breakers have a somewhat higher fire hazard in the event of a fault that may vaporize sufficient oil to rupture the enclosure. Circuit breakers serve both as a switching device and as a protective device to interrupt the flow of excessive currents resulting from short circuits or overload. Circuit breakers are identified by the normal current and voltage that the breaker can safely handle as a switching device. The interrupting capacity of circuit breakers is the current that the circuit breaker is designed to interrupt for protective purposes. The calculation of the short-circuit interrupting capacity of circuit breakers should be left to those competent in the electrical field and depends upon several factors, including the electrical load of the circuit, interrupting capacity of other circuit breakers in the circuit on the supply side of the breaker, and others.

Wiring

While the installation design of wiring is again left to electrical engineers, consideration must be given to the location of feeder circuits with regard to possible interruption of production by a fire and by poor location and poor protection of wiring. Power lines should preferably not extend across the roofs of combustible areas and should be located so as not to be exposed to outage, in the same fashion that switch gear and transformers would be located with regard to special hazards. This is a particularly true of feeders for fire or booster pumps. It is desirable to use rigid metal conduit for important feeders wherever possible.

Of special consideration also from the standpoint of continuity of operations is the grouping of cables in cable troughs or cableways there fires can occur from faults that cause feedback from the closely grouped cables. These cables are normally from control rooms in large plants or master control rooms for television broadcasting where an interruption by a small fire of this nature can be disastrous. Grouped cables of this type should protected by a row of automatic sprinklers; in lieu of this, it may be advisable to provide a continuous-type detection cable that can be instrumented to indicate within a relatively few feet the location of a high-temperature condition. Commercially available cable coatings can be used to prevent the spread of flame without causing undue heating of the conductors. These coatings should be applied to vertical runs at changes in direction from the horizontal to vertical and also where these cable trays extend through wall or partition openings. Coatings that will not cause a derating problem should be used.

Heating, Ventilating, and Air Conditioning

Improper setting of heating appliances, chimney and venting systems, and the lack of adequate combustion safeguards have resulted in numerous fires and, in the case of combustion safeguards, firebox explosions. Also, the arrangement of duct systems, the use of combustible insulations and sound liners, and the lack of dampers have allowed the passage of smoke and heat throughout buildings. The topic of cooling towers also has a place in our discussion under this general topic, because of their combustibility, but even more so for the effect that the loss of towers has on the continuity of operations.

Heating Systems

In addition to consideration for proper setting of heating appliances with relation to combustible materials, it is also important to arrange for the safe removal of the products of combustion and the heat produced using safe chimneys or venting systems. The *Standard for Chimneys, Fireplaces, Vents, and Solid Fuel Burning Appliances*, NFPA (2), describes chimney and venting systems and provides minimum acceptable clearance to combustible materials under varying conditions. An example of information found in the standard is the minimum clearance of 18 in. for single-wall metal pipe connectors to combustibles for gas, oil, and solid fuel boilers, furnaces, and water heaters.

Fuel explosions in fireboxes, while relatively infrequent, can create extensive property damage. Many of the explosions also create problems of business continuity or interruption, possibly resulting in losses greater than the physical losses. Explosions of this type occur in an explosive mixture of fuel and air with an ignition source present. The ignition source may be the normal ignition source for the fuel, such as a pilot, or may be from other sources within the firebox or passageways, such as hot refractories after a loss of flame.

Normally, with small-capacity heating units or furnaces, it is not anticipated that a loss of flame would produce a serious explosion in the combustion chamber, even with slow-reacting thermal cutoff controls such as stack switches, because a small quantity of the fuel was released into the chamber before fuel shutoff. In large installations, however this is not the case: slow-reacting combustion controls such as the thermal type permit large accumulations of raw fuel in the firebox prior to shutoff, which could result in significant fuel explosions. For this reason, more sophisticated combustion safeguards are required for larger furnaces or boilers, including nominal 3-sec flame safeguard controls.

Most large-risk underwriters require faster acting, additional, and more reliable controls for the larger boilers. For units having input ratings of 400,000 Btu/hr or less, slower acting controls and fewer interlocks are permitted. These include American Gas Association (AGA) or UL approved controls. The flame-sensing unit and the safety shutoff valves are considered the key units in a combustion safeguard system. Flame detectors or sensors are designed as nominal

3-sec type, which means that they will respond within 2 to 4 sec after flame failure. Upon sensing a loss of flame, they will cause the safety shutoff valve to close before permitting a dangerous accumulation of fuel within the firebox.

The proper application of flame detection is important and the following general comments are presented, although they are by no means to be considered a complete covering of this topic. Lead sulfide cells, commonly used for both fuel oil and gas, detect infrared radiation. This type of sensor can be "fooled" by hot refractories that also give off infrared radiation. Ultraviolet-type detectors are not "fooled" by hot refractories but can, however, detect ultraviolet rays from electric sparks such as from electric ignition, which could result in the premature opening of the main fuel valve.

Ultraviolet detectors will also cause a shutdown if a haze exists that prevents the ultraviolet rays from reaching the detector. Cesium photocell detectors or sensing fuel oil flames only (gas flames are not luminous). Flame rods use the flame rectification principle to detect flames, are relatively simple in installation, and are also fail-safe. Rods can be somewhat difficult to get into the flame and also present the possibility of causing shutdowns if touched by metal.

Safety shutoff valves must have a high degree of reliability because these valves must close tightly (and quickly) when an unsafe condition is detected. Approved valves are held open by electric energy or pressure, which may be air or gas, and are closed by the action of a deadweight or spring. Approved valves must close within 5 sec after the holding medium is cut off. These valves are enclosed so that they cannot be bypassed or blocked in an open position. With a manually opened valve, blockage of a handle will not prevent the closing of the valve when the holding medium is cut off. Safety shutoff valves, if used in conjunction with another safety shutoff valve, can be used as operating valves.

These valves, as well as any other type of valve, are known to leak occasionally; consequently, leakage tests should be conducted periodically. For this purpose, petcocks are provided downstream of the valves, permitting testing with the valve closed by opening the petcock into a tube discharging from the petcock through water. Obviously, a leaking valve will result in gas bubbling through the water.

Because of the possibility of leakage, particularly with large-rated burner assemblies, it is desirable to provide two valves in the main train with the second valve designed as a block valve. By providing a vent to the atmosphere between the two valves with a reverse-acting valve to be normally open when the two main valves are closed, any gas leaking past the first valve will be vented to the atmosphere rather than to the firebox, which would occur with a single valve in the gas train. A gas train with this venting arrangement is shown in Figure 8.6.

For gas burner ratings below 2,000,000 Btu/hr input, underwriters may permit the use of an operating valve as one of the blocking valves in place of a safety

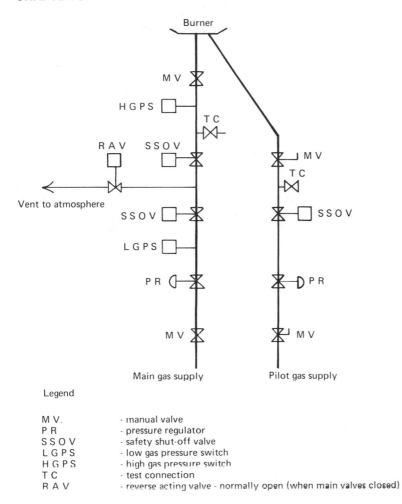

Figure 8.6 A typical piping schematic for combustion safeguards. Note: Gas pressure switches and pressure regulators may use a common vent to the atmosphere.

shutoff valve. In this arrangement, the preferable location of the operating valve is on the downstream side of the vent.

Combustion systems often depend upon air for proper combustion or atomizing. The loss of this medium can create an unsafe condition; consequently, a combustion safeguard system should include a means of supervising air such as an airflow switch.

Stable flame is dependent to a great degree, particularly in the case of natural gas, on proper fuel pressures. A high gas pressure outside safe limits can result in the blowing-off of the flame from the burner tip, with a low fuel pressure also resulting in a possible flameout. Both conditions can result in the pumping of raw fuel into the firebox prior to the activation of the shutoff valve. To prevent situations of this type, high-and low-gas-pressure switches and interlocks are provided to prevent operations outside of safe limits. Similarly, low fuel oil pressures can have the same consequences, and a low pressure oil switch and interlock is desirable in oil lines feeding burners.

Proper viscosity is also required in the case of fuel oil, which can be ensured by providing a low-temperature interlock.

Prepurging of the firebox and passages should be done to prevent accumulations of raw fuel at start-up or following a shutdown. Four air changes prior to ignition are generally considered adequate.

With large burner inputs, it is necessary to restrict the time that a safety shutoff valve may be open before the combustion safeguard is required to prove a flame at both the gas pilot and main burner. In the case of gas pilot, this is known as the pilot flame establishing period, and with a main burner, the trial for main flame. The trial for main flame with gas and light fuel oil should be limited to a maximum of 15 sec. With heavy fuel oils, a maximum of 30 sec may be permissible. The pilot flame establishing period should be restricted to 10 sec.

From the preceding paragraphs, it is evident that a good design of the combustion system will provide

1. A proper prepurge prior to the proving of both the pilot and main burner flames
2. The detecting and interlocking of fuel unsafe conditions such as pressure or temperature
3. The supervising and interlocking of combustion air
4. Trial periods for both pilot and main flames
5. Lockout on flame failure

Approval authorities, when reviewing combustion safeguard submittals, prefer the wiring schematic in a "ladder" form; a sample is shown in Figure 8.7. A piping diagram depicting the arrangement of the combustion safeguard components and a list of the components by model number and manufacturer are also

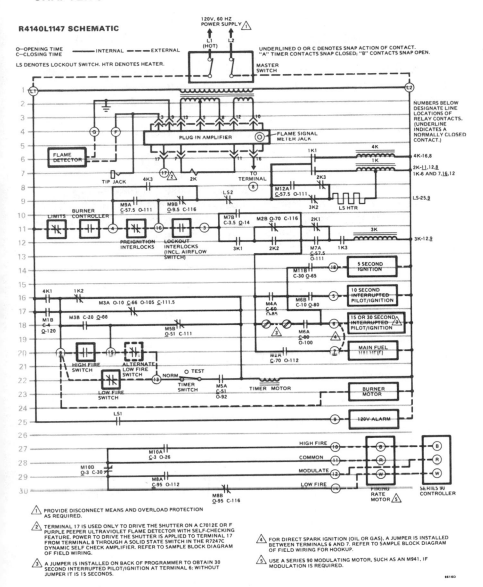

Figure 8.7 Wiring schematic in "ladder" diagram form of a flame safeguard programmer that includes the burner controls. (Courtesy of Honeywell, Inc.)

required. An example of a piping diagram is shown in Figure 8.6. Only approved controls should be used. Consult the UL *Oil and Gas Equipment List* (23) and the FM *Approval Guide* (13) for these listings.

Where multiple burners are involved, it is desirable to provide FM cocks (supervisory cocks) to ensure that the valves to each burner are closed prior to opening of the safety shutoff valve. Using the supervisory cock, all valves to burners must be closed to permit the flow of gas used as a checking medium to close a checking pressure switch permitting the main gas valve to open.

Air-Conditioning and Ventilating Systems

The *Standard for the Installation of Air Conditioning and Ventilating Systems,* NFPA 90A (2), covers air-conditioning and ventilating systems. The basic concern is the transfer of smoke through a duct system and, to a somewhat lesser extent, the passage of gases and the actual fire. This concern is heightened in multistory, particularly high-rise buildings, where the effects of smoke may be devastating.

The principal objectives in good design of air-conditioning and ventilating systems are

Avoid any combustibles within the duct system, including filters, duct liners, and
   the duct construction itself.
Avoid combustible exterior duct insulation.
Prevent the passage of smoke and fire through the ducts.
Maintain the integrity of fire and fire-barrier walls where penetrated by ducts.
Maintain the integrity of floors where penetrated by ducts or connectors.

The first objective is accomplished by the use of noncombustible duct material and insulation. UL tests and lists ducts and classifies them as follows:

Class O: Ducts and connectors with surface burning characteristics of zero.
Class 1: Ducts and connectors with flame-spread rating not over 25, without continued progressive combustion, and smoke-developed rating not over 50.
Class 2: Ducts and connectors with flame-spread rating not over 50, without continued progressive combustion, and smoke-developed rating not over 50 for inside surface and not over 100 for outside surface.

Because the use of glass fiber-reinforced polyester ducts where high corrosive atmospheres are present may pass fire through the shielded combustible material, in-duct sprinkler protection is advised. Additives such as antimony trioxide have been used in these ducts to reduce combustibility, but without a great deal of success. This has emphasized the need for sprinkler protection in larger ducts (12-in. diameter or larger). For smaller diameter ducts, polyvinyl chloride and glass-reinforced polyester type should be acceptable under sprinkler protection. It may be desirable to provide access to the duct interior for hose stream insertion. These ducts should not be installed above ceilings without sprinkler protection, and larger diameter ducts should be provided with in-duct sprinklers. Corrosion protection of heads and piping is necessary where highly corrosive atmospheres may be encountered.

Duct covering and lining materials should have a maximum flame spread of 25 without evidence of continued progressive combustion and a maximum smoke-developed rating of 50. Where ducts penetrate floors or walls, linings and insulation should be interrupted and also where their use interferes with the operation of fire doors or dampers.

Figure 8.8 depicts a typical air-conditioning system in a fire-resistive multistory building. This shows clearly the requirement of enclosing ducts that pierce floors in 1-or 2-hr fire resistance rated shafts with 1-hr required for buildings having four or less stories. In other words, these are vertical openings and must be treated as vertical openings. An exception to the rule for enclosures is where fire dampers are provided at floor openings where ducts that are located in one story extend only to either the floor above or below. Fire dampers are required where a duct passes through fire barriers of 2-hr rating or less and at each opening in a required enclosure of a vertical shaft with some specific exceptions as listed in NFPA 90A (2).

Approved fire dampers are listed in the UL, *Building Materials Directory* (10).

Air-conditioning systems can be used to exhaust smoke under fire conditions when carefully engineered for that purpose. This may involve the air-conditioning system in conjunction with emergency venting, pressurizing, and fire suppression. From a life safety standpoint, this arrangement should be beneficial for buildings having zones or stories required to be separated by construction that will restrict the spread of smoke or fire and where evacuation time could be excessive or not practical due to the occupants (age, physical and mental disability, or security measures). Guidelines for this type of system can be found in NFPA 90A (2).

In systems not designed to change to an exhaust mode under emergency conditions, it is usually desirable to shot off airflow. NFPA 90A (2) requires smoke detectors approved for duct installation in systems over 15,000 ft^3/min arranged to automatically shut down the fans and close smoke dampers. Detection should be located upstream of filters and prior to dilution with any make up air. Systems having capacities between 2000 and 15,000 ft^3/min, should be equipped with fixed-temperature thermostatic devices with a setting not exceeding 136°F (58°C) located in the return airstream prior to dilution by makeup air and with a setting not in excess of 50°F (10°C) above the maximum operating temperature, and located in the main airstream on the downstream side of the filters. The op-

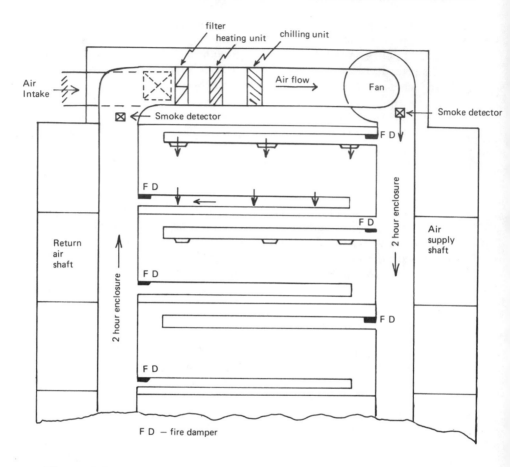

Figure 8.8 A typical air-conditioning system found in multistory buildings.

eration of these thermostatic devises in high-velocity airflows is highly questionable.

Filters of either approved Class 1 or Class 2 as listed in the UL *Building Materials Directory* (10) should be used. In large systems in excess of 20,000 ft ^{3}/min, it is advisable to provide automatic sprinkler protection at filters with provision for drainage.

Cooling Towers

Water-cooling towers are frequently a highly vital building component in ensuring continuance of operations by providing cooling water for air conditioning, refrigeration, air compressors, and other equipment using water for cooling. Even if not involved in processing, cooling towers for air-conditioning systems in large facilities (for example, a large electronic assembly plant) are frequently essential to maintain operations. These towers can be of considerable size and, because most are of combustible construction, represent a loss potential, although mostly wetted. In most towers, the fill portion is of wood and burns readily when dry. Also, in many towers, fan decks and other portions of the tower are not wetted and are exposed to fire even when in operation. The sources of ignition of towers may be sparks from nearby chimneys or stacks, from welding or cutting operations, or from other sources of ignition such as smoking and electrical faults. The *Standard on Water-Cooling Towers*, NFPA 214 (2), covers this topic and should be consulted for details of good location, design, and protection.

Towers are of two basic types, atmospheric and mechanical draft. Atmospheric types depend upon the natural movement of air for cooling and normally represent no problem when in operation. Large hyperbolic types are used by utilities.

Mechanical draft towers are usually of more concern in fire protection from the standpoint of size, value and the need for protection. This class of tower can be further broken down into induced-draft, counter-or cross-flow, or forced-draft types. Of these, induced-draft, counter-and cross-flow types are more common with normally combustible fan decks as well as fill.

The location of towers should be given careful consideration to protect against exposure hazards that may involve highly flammable and volatile materials or combustible debris or weeds. When ground supported, the tower or towers should be fenced and sufficient distance provided between exposing hazards or combustible construction.

The main problem of tower combustibility can be bypassed by the use of noncombustible fill rather than redwood type. For existing towers, a major firm has developed a fire-retardant coating that will resist the leaching action of the water and can be applied to the combustible fill, thereby reducing combustibility.

Protection of towers is usually found in combustible towers when justified by the value or need to maintain continuity of operations.

Protection in the case of induced-draft, counter-flow towers will take the form of wet-or dry-pipe, closed-head sprinkler systems or deluge (open-head) systems. Where towers are large and water supplies limited, closed-head systems are preferable. Open-head deluge systems are normally required for cross-flow type. NFPA 214 (2) should be consulted for details, including recommended design densities. Where water supplies are available, hydrant protection should be provided for large towers within 200 ft. Combustible towers on building roofs should also be proceeded by hose streams from standpipes.

One last not pertaining to water towers and the need or lack of need for protection is given here. For installations that must remain in operation, it may be feasible to provide for hose connections from a municipal supply for emergency periods when the tower is out of operation. This obviously will increase costs for an interim period and will require availability of city water through hose and preplanning for this possibility.

SPECIAL FIRE PROTECTION CONSIDERATIONS

Static Electricity

The generation and subsequent control or removal of static electricity is of considerable consequence in fire protection engineering where flammable vapor mixture or combustible dust may be present, such as in operations involving the mixing or flow of flammable liquids or the handling or processing of dust.

The generation of static electricity in itself may or may not be dangerous unless certain conditions exist:

1.   There is a concentration of flammable vapors or dust within the flammable limits.
2.   The charge generated is not dissipated.
3.   The static spark or discharge is of sufficient intensity or energy to cause ignition.

To generate static electricity, there must be an interface and a separation of two dissimilar materials, one of which is nonconductive. This can happen when a liquid flows through air, such as when a flammable liquid is transferred from one container to another; when a tank is filled and the fill pipe is not brought into proximity with the bottom of the tank; when various types of dusts pass through ducts or chutes; or when nonconductive conveyor belts are in motion. If one of the materials in contact is conductive, there is no buildup of charge on this material; however, if the other material is nonconducting or dielectric, static charges can be accumulated until the energy level reaches the point where a spark will be produced. Experience has shown that the flow of flammable liquids through closed piping systems presents no static hazard, with bonding required only around joints in which the only contacting surfaces are of nonmetallic insulating material. In general, gases generate little, if any, static unless they contain particulate matter such as metallic oxides or scale particles. Gas in a closed piping system presents no particular hazard, and these systems need not be electrically conductive or bonded.

Operations involving combustible materials such as a flammable vapor or dust, when in contact with or separated from dissimilar materials do require static electricity protection. Recognizing that it is virtually impossible to prevent the generation of static electricity, attention must be directed to dissipating charges. This is generally done by grounding or bonding, although other means are also available for more specific situations.

There often appears to be confusion between the terms *bonding* and *grounding*. Grounding is the bringing of a metallic object into contact with the ground such as by a wire conductor attached to a flammable liquid tank with the wire then attached to a metal rod driven into the earth. Bonding is the connecting of two containers, two sections of pipe, etc., by an electrical conductor to minimize the difference in electrical potential (voltage) between the two bodies and prevent a possible discharge of static between them. The terms are closely related and, in effect, grounding is a form of bonding. Figure 6.5 in Chapter 6 illustrates the difference between grounding and bonding, showing both the bonding of a drum to a safety can be filled and the grounding of a drum. Bonding or grounding wires, because of the small amounts of current carried, are not required to be of extremely low resistance, but should be of adequate capacity to carry the largest anticipated currents. Mechanical strength is generally more of a consideration when expected current capacity is small. Flexible conductors are desirable for bonds that are connected and disconnected, to prevent frequent breakage.

For the ground or bond connections to perform their function, there must be adequate metal-to-metal contact. Often, clamps or alligator clips are affixed to drums or other containers that have a heavy coating of paint or other material that acts as an insulator, preventing the dissipation of any charges. This is particularly true of paint plants.

Other methods of removing static electricity are humidification and ionization. The conductivity of any material, even a nonconductive one, can be improved to dissipate static charges by providing a film of moisture over the surface. Conductivity is related to relative humidity, and a 70% relative humidity is usually required to provide dissipation of charges. Because this high humidity is generally unsatisfactory for most situations, humidification is a safe means of static removal only in a very few situations. Operating rooms in which combustible anesthetics are used are generally maintained at 50% to 55% humidity, which, although assisting in the dissipation of static electricity, is not considered absolutely safe. In these instances, further measures are taken in the form of conductive floors, casters, conductive-soled footwear, etc.

Ionization makes the air adjacent to static eliminators sufficiently conductive to dissipate static charges. This is done through the use of grounded metal bars with needle points or "combs" close to a charges insulating surface. This method is often used to remove static charges from fabrics or paper being coated or impregnated, particularly with the use of roller-type coaters, and also to remove charges from power belts.

For a detailed treatment of the subject of static electricity, refer to the *Recommended Practice on Static Electricity*, NFPA 77 (2).

Data-Processing Equipment

In modern society, the computer has become an extremely important tool to business and industry in conducting their activities, in providing rapid accounting functions, in providing process and inventory control, and in other ways. Consideration should be given to providing protection for data-processing installations in view of the high values found in these areas, the susceptibility of the equipment, and from the disruption to business continuity that could result from the loss of data-processing facilities. The degree of concern and protection required will depend upon an analysis of several variables, including the following:

1. The value in dollars of the equipment, and ownership and responsibility for the equipment.
2. The functions of the equipment. Is the installation involved vital for maintaining important functions, and is backup to the equipment and media readily available?
3. Any uniqueness involved in the equipment.
4. The exposures to the equipment, both external to the computer enclosure (frequently termed the *envelope*) and within the data-processing area itself.

The first consideration in the data-processing installation is building construction and the location of the data-processing area.

These comments pertain primarily to fire protection and related areas and do not include a consideration of the security aspects. The *Standard for the Protection of Electronic Computer/Data Processing Equipment*, NFPA 75(2), states that the "electronic computer area shall be located to minimize exposure to fire, water, corrosive fumes, and smoke exposure from adjoining areas and activities."

The computer room should be suitably enclosed to protect it from adjacent exposures, even if these are relatively low-hazard office exposures. In this case, it is suggested that a minimum 1-hr fire-rated enclosure be provided with openings protected accordingly. If the exposure has a greater potential fire loading than does the enclosure, the fire resistance rating should be raised commensurately.

Often, the location of the envelope is not given sufficient thought, so that the area may be subject to water conditions, which could be the case with a basement location or with a location beneath processing involving water. The location of an extremely valuable and vital data-processing operation subject to water conditions or unusual fire hazards should be strongly discouraged.

Once the fire resistance of the enclosure, which can be of relatively light construction to meet the fire rating required, has been determined, selection of the interior finishing should be made, and the interior finishing should have a flame-spread rating of 25 or less. This would include a raised floor. Even though it is preferable to have a hard surface covering, carpeting is now being used in data-processing areas on the raised floor. If carpeting is used, it is recommended that the carpeting have a flame-spread rating of 25 or less and not restrict access to space under the floor.

Tape and paper should be stored in rooms cut off from the main data-processing equipment. Tapes are normally stored in plastic containers, which represent a high hazard exposure to the data-processing equipment in the form of high heat release of energy and also the generation of considerable smoke.

Computer equipment and materials for data recording and storage are susceptible to elevated, sustained temperatures. Data (magnetic tapes, flexible disks, etc.) will be damaged at temperatures in excess of 100°F (38°C). At temperatures of about 175°F (79°C), damage will be done to the equipment itself. Smoke also can create erratic behavior of the equipment.

The necessity and type of protection for the computer envelope depend to a great extent upon the ability to control or limit the quantity of paper and similar materials. If firms are able to operate with minimal material of this kind within the envelope, sprinkler protection is not needed, assuming the absence of combustible construction. Conversely, if the quantity of combustible materials cannot be controlled, as is the case in numerous installations, sprinkler protection is advisable. To overcome the possibility of accidental water discharge, it is suggested that a preaction sprinkler system be considered, although it is not felt that accidental discharge has been of any consequence in data-processing installations. One sprinkler head manufacturer has developed an on/off sprinkler head that is particularly suited for data-processing installations. It reduces the quantities of water needed because the head has the ability to turn off at 100°F (38°C). Early models of this head have leaked in some instances, but the latest model appears to have overcome this problem. In addition, to minimize damage to computer equipment, a means should be provided to disconnect the power to the equipment prior to sprinkler discharge.

Although the envelope itself may be sprinklered, the underfloor area and the cabinets housing data-processing equipment are shielded from the sprinkler discharge. Consequently, underfloor automatic extinguishing systems should be provided where considerations of high values or continuity of operations are involved and where there is extensive grouping of polyvinyl chloride and polyethylene-insulated conductors not in conduit (cabling normally exposed for flexibility). Halon 1301 systems of a total flooding nature have in the past provided protection of the entire computer envelope, but due to the limits being placed on Halon use and production (see further details in Chapter 10), other suppression alternatives to Halons may have to be found in the future. The extinguishing system will also totally flood the interior of the cabinets, since in most data-processing installations, the underfloor area is used as a plenum with discharge up through the boxes to exhaust louvers in the ceiling.

Because the equipment is highly susceptible to smoke damage, smoke detection should be provided both at the ceiling of the computer envelope and beneath the raised floor. In consideration of the airflow, closer spacing of detectors is advisable. In cases where a fire may operate sprinkler heads before discovery by personnel, it is also advisable to arrange for the detection system to disconnect power.

In addition to the fixed protection systems mentioned above, portable carbon dioxide extinguishers should be provided for discharging into the computer cabinets or beneath the raised floor area. In addition, there should be also Class A-type extinguishers available for handling fires in ordinary combustible materials.

While it is recognized that there may be opposition to the installation of automatic sprinkler protection within computer rooms, it should be emphasized that this is the only feasible way to provide protection where there are extensive quantities of Class A materials. With the installation of automatic sprinkler systems, waterproof covers should be provided for the more sophisticated and valuable pieces of data-processing equipment. In addition to delay discharge of water from sprinkler heads, the temperature rating of heads may be higher, thereby giving additional time for the possibility of suppressing fires manually.

With regard to air-conditioning systems, which are vital to the continued operation of data-processing equipment, separate air-conditioning should be provided for the computer room as opposed to the use of general building systems. Good practices in the installation of ducts and in providing noncombustible insulation and duct liners should be followed. The system should be provided with duct smoke detection located upstream of the filters, which will disconnect the air-conditioning system, particularly as indicated above, if an automatic gas system for extinguishment is provided beneath the floor or, for that matter, in the entire computer envelope.

It cannot be overemphasized that preplanning in the establishment of emergency procedures is essential for data-processing areas. Preplanning includes the arrangement for storage of data either off premises or in a section that will not be involved in a catastrophe affecting the data-processing area. Also, backup data-processing capabilities should be investigated and arrangements made for the use of similar data-processing equipment employing the software programs, which should also be in duplicate and located off premises. The emergency planning should provide for the manual shutdown of data-processing equipment, if necessary, the removal of vital records, the notification of the fire department, the use of portable extinguishers, etc.

SUMMARY

Hazard evaluation and provisions for adequate fire protection should be an integral part of the basic engineering design for all new installations, alterations, or additions. Frequently, minor changes can effect major savings. For example, problems encountered with one raw material might dictate substitution with another that is safer to use but equally effective. Relocation of processing equipment, rearrangement of barrier walls, or changes in flow patterns can often be made readily and with minor expense in the design stage. However, once built, the elimination or protection of hazards either overlooked or ignored in the original planning can be considerably more costly

It is the mark of good management and competent engineering to make certain that hazard analysis and provision for adequate fire protection are required considerations in every phase of plant design and construction.

REFERENCES

1. Cote, Arthur E., ed., *Fire Protection Handbook*, 16th Edition, National Fire Protection Association, Quincy, Mass. 1986.
2. *National Fire Codes*, National Fire Protection Association, Quincy, Mass. 1989.
3. Nelson, Harold E., *The Application of Systems Analysis to Building Fire Safety Design*, Accident and Fire Prevention Division, General Services Administration, (undated).
4. Lerup, Lars, Cronwrath, David, and Liu, John Koh Chiang, *Learning from Fire: A Fire Protection Primer for Architects*, U.S. Fire Administration, Washington, D.C., 1977.
5. "Flood! And Why You're Not as Safe as You Think," P 8001, Factory Mutual Engineering Corporation, Norwood, Mass., 1980.
6. "National Disasters," P 8812, Factory Mutual Engineering Corporation, Norwood, Mass., 1988.
7. "Earthquake Concerns," P 8805, Factory Mutual Engineering Corporation, Norwood, Mass., 1988.

8. *Special Technical Publication 464*, American Society for Testing and Materials, Philadelphia, Pa., 1970.
9. *Fire Resistance Directory*, Underwriters Laboratories, Inc., Northbrook, Ill., 1989.
10. *Building Materials Directory*, Underwriters Laboratories, Inc., Northbrook, Ill., 1989.
11. *Standard Fire Tests of Building Construction and Materials*, UL 263, Underwriters Laboratories, Inc., Northbrook Ill.,
12. *Loss Prevention Data Books*, Factory Mutual Engineering Corporation, Norwood, Mass., 1989.
13. *Approval Guide*, Factory Mutual Engineering Corporation, Norwood, Mass., 1989.
14. *Standard Test Methods for Higher Hazard Classifications of Building Materials*, UL 723, Underwriters Laboratories, Inc., Northbrook, Ill.,
15. Waterman, T. E., "Fire Venting of Sprinklered Buildings," *Fire Journal*, 78 (2): 30–39, 86 (1984).
16. Heskestad, Gunnar, "Model Study of Automatic Smoke and Heat Vent Performance in Sprinklered Fires," *Technical Report FMRC Serial No. 21933 RC 74–T–29*, Factory Mutual Research Corporation, Norwood, Mass., 1974.
17. *The Foundations of Loss Control*, Factory Mutual Engineering Corporation, Norwood, Mass., 1987.
18. Butcher, E. G. and Parnell, A. C., *Smoke Control in Fire Safety Design*, E. and F. N. Spon, London, England, 1979.
19. Runes, E., "Explosion Venting," *Loss Prevention*, 6: 63–67 (1972).
20. *General Industry Standards—29 CFR 1910*, Occupational Safety and Health Administration, Washington, D.C., 1988.
21. *Hazardous Location Equipment List*, Underwriters Laboratories, Inc., Northbrook, Ill., 1989.
22. Earley, Mark W., ed., *The National Electrical Code—1990 Handbook*, National Fire Protection Association, Quincy, Mass., 1990.
23. *Oil and Gas Equipment List*, Underwriters Laboratories, Inc., Northbrook, Ill., 1989.

9
Water Supplies for Fire Loss Control

As stated in Chapter 8, Fire Safety in Design and Construction, one of the main considerations in plant site selection is the determination of the availability and strength of the water supply. Water is the basic agent used to extinguish building fires. While other fire-suppression systems are used, water remains the primary agent used by the fire department. With this in mind, the designer needs to look at the water supply requirements of the fire department and the automatic suppression systems in the facility to provide the necessary supply at an adequate pressure. This information can often be obtained from municipal officials and from a study of the municipal water distribution system, with the important inclusion of water test data at the plant site. Test results are then subject to analysis to determine the quantities of water available at various pressures and the reliability of the system.

TESTS OF WATER SUPPLIES

In conducting tests of water supplies, a simple instrument known as a pitot tube is used by fire protection engineers. This device is inserted into water streams to measure the velocity, which is converted to total head in pounds per square inch (psi) on a pressure gauge attached to the tube. Typical pitot tubes and accessories available commercially are shown in Figure 9.1. Specifications of a typical pitot tube and assembly and hydrant cap and gauge are shown in Figure 9.2. The flow in gallons per minute can then be calculated or, more conveniently, obtained from

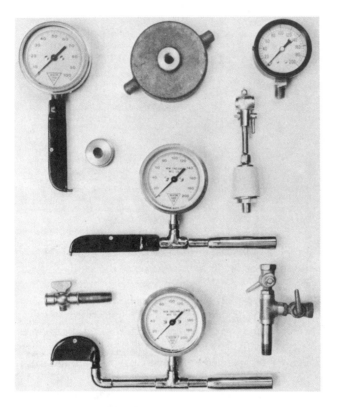

Figure 9.1 Typical pitot tubes and accessories used in water testing. (Courtesy of New England Manufacturing Co.)

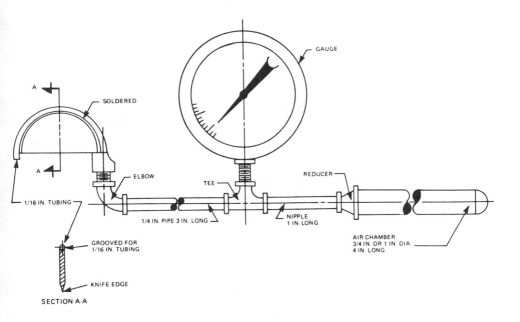

Figure 9.2 Pitot tube and assembly and hydrant cap and gauge. (Reprinted with permission from the *Fire Protection Handbook*, 16th Edition, Copyright 1986, National Fire Protection Association, Quincy, MA 02269.)

tables listing discharges for circular outlets, which are available in various publications, such as NFPA *Fire Protection Handbook* (1), FM *Loss Prevention Data Books* (2), American Mutual Insurance Association's *Water Supply Testing* (3), and American Insurance Association's *Fire Flow Tests* (4). An example is shown in Table 9.1.

In regard to the pitot tube generally used to obtain water test data, the reader should be aware of the general method of conducting water tests, which involves flow from open hydrant butts or, to obtain somewhat more accurate results, laying hose from hydrants and measuring flow from nozzles (called *underwriter playpipes*). When using hose and nozzles, it is necessary to convert the residual pressure at the hydrant to nozzle pressure by considering the flow through the hose. Tables for determining the discharge from nozzles attached to the hose, considering the friction loss, are available in various hydraulic publications.

In any test, the exact diameter of the nozzle or orifice discharging water must be obtained and accurate gauges must be used. Because gauges are sensitive, they should not be handled roughly and should be recalibrated frequently. While one

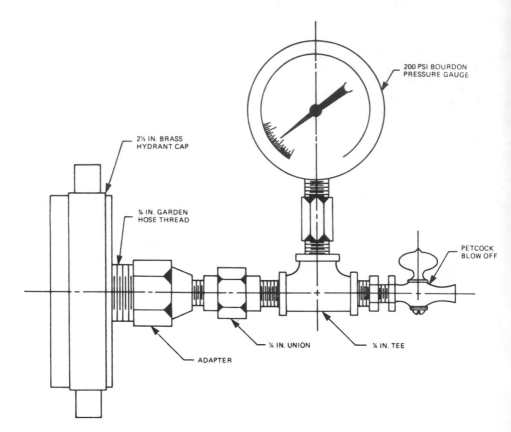

Figure 9.2 Continued

water test, with the flow determined at a particular residual pressure and the static pressure known, can enable the plotting of the water supply on exponential paper, it is desirable to conduct tests at two or more flows for greater accuracy.

In making flow tests, flows should not reduce the pressure in mains to below 20 psi to avoid possible collapsing of mains. Caution should also be exercised in the opening and closing of valves to avoid damage from water hammer, which will require careful and slow operation of all valves.

When testing on the premises of a plant, to obtain more steady flows and more accurate results, water tests are usually made by attaching the hose lines to the hydrants and taking readings at the nozzles. Tests should be made at flows that nearly approximate the water demand of the plant. Tests should be made in the

Table 9.1 Discharge Table for Circular Outlets[a]

Pitot Pressures (psi)	Nozzle Diameters			Pitot Pressures (psi)	Hydrant Butt Diameters			
	1"	1 1/4"	1 1/2"		2 7/16"	2 1/2"	2 9/16"	4"
6	72	113	163	6	434	457	480	1170
8	84	131	188	8	501	528	554	1351
10	93	146	211	10	560	590	620	1510
12	102	160	231	12	614	646	678	1655
14	110	173	249	14	664	698	732	1787
16	118	185	267	16	709	746	783	1910
18	125	196	283	18	752	792	832	2026
20	132	206	298	20	793	834	875	2136
22	139	216	313	22	832	875	918	2240
24	145	226	327	24	869	914	960	2340
26	151	235	340	26	903	951	1000	2435
28	157	244	353	28	938	987	1036	2527
30	162	253	365	30	970	1022	1072	2616
35	175	273	394	32	1003	1055	1108	2702
40	187	292	422	34	1034	1088	1142	2785
45	198	310	447	36	1064	1119	1175	2866
50	209	326	472	38	1092	1150	1207	2944
60	229	357	517	40	1120	1180	1239	3021
70	247	386	558	42	1148	1209	1270	3095
80	264	413	596	44	1175	1237	1298	3168
90	280	438	633	46	1202	1265	1329	3239
100	295	461	667	48	1228	1293	1357	3309
				50	1254	1319	1385	3377
	coefficient = 0.99				coefficient = 1.0			

[a]Reprinted with the permission of American Mutual Insurance Alliance.

vicinity of the area having a maximum water demand or, perhaps, the highest concentration of values. In a great many cases, these two locations will be the same.

Water tests made from public mains almost invariably involve taking pitot readings from open 2.5-in. hydrant butts; again, it is desirable to have two or more tests made for accuracy purposes.

In conducting the test, water is flowed through one hydrant (whether a hose is attached to it or not) with static and residual pressures observed on a nearby non-flowing hydrant. In plotting a water supply, it is necessary to record both static and residual pressures. For those not familiar with hydraulic terms, pressure observed with no flow or, in essence, the pressure exerted by the system when water is at rest is termed *static pressure*. Pressures obtained at the nonflowing hydrant or at other gauges, such as a sprinkler riser gauge with water flowing through the flowing hydrant, are termed *residual pressures*. Static pressure in itself is of little consequence, because it has no relationship to the actual quantity of water available. It is, however, significant in plotting a water supply.

When making the actual water test, residual pressures should be recorded at the hydrant nearest the plant site or main value concentrations. Flow should be made from the next nearest hydrant, which, in the event of a one-way feed, should be farthest from the water supply.

The actual quantity of water flowing is determined from the pitot reading, which actually measures the velocity pressure of the stream. The pitot tube is held firmly in the center of the water flow just outside the butt or nozzle with the blade at right angles to the nozzle axis. Obtaining the velocity pressures, or pitot readings, the corresponding flow in gallons per minute is obtained from tables similar to Table 9.1. Pitot readings under 10 psi and over 30 psi are very questionable. Readings under 10 psi should be avoided by reducing, wherever possible, the size and number of flowing orifices. Readings over 30 psi should be avoided by increasing the number of flowing on orifices.

Exponential graph paper is used to clearly illustrate the results of water test data, showing graphically the capabilities of combined supplies as well as the friction loss in pipelines at various flows. This paper will normally plot test flows as straight lines, which would not be the case in the use of linear-cross-section paper. With the use of exponential (to the 1.85 power) paper, knowing the static pressure and the flow at a particular residual pressure, a line can be drawn that will show the flows available at the test point at any pressure. This is illustrated in Figure 9.3 with an example that also shows how two supplies can be obtained to show their joint capability.

An example follows of a flow test using street hydrants to determine the water supply for an industrial connection.

Conduct test as follows (refer to Figure 9.4):

1.   Attach gauge to hydrant no. 1. Open hydrant and record static pressure (70 psi recorded). Note: 10-in. main branches into 8-in. main in front of the plant.

2.   Choose flow hydrant closest to plant connection. Remove cap from hydrant no. 2, measure diameter of outlet (2.5 in. measured), and determine type of outlet (outlet determined to be smooth and rounded). Open the hydrant, take a pitot tube reading, and record pressure (20 psi recorded). While hydrant is flowing, record residual pressure at hydrant no. 1 (60 psi recorded). Shut down hydrant no. 2 slowly.

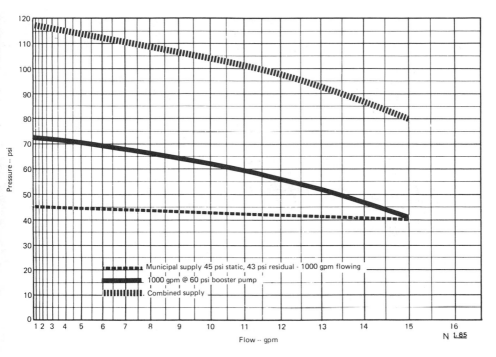

Figure 9.3 The plot of water supplies on semiexponential paper ("hydraulic" graph paper) showing two supplies and their joint capability. (Note: flow scale = × 100 gal/min.)

3. Remove second outlet cap on hydrant no. 2, open hydrant, take pitot reading on both outlets, and record pressures 14 psi recorded on each outlet). Record the residual pressure at hydrant no. 1 (42 psi recorded). Note: If possible, obtain data for at least two rate of flow.

4. Shut down hydrant no. 2 slowly and replace caps. Recheck static pressure at hydrant no. 1 (70 psi). Shut down hydrant, remove gauge, and replace cap.

Compute the flow: Using a table of theoretical discharge (Table 9.1), for pitot pressure of 20 psi, flow from 2.5 inch outlet is 834 gal/min. Actual flow is 834 × 0.9 = 751 gal/min. at residual pressure of 60 psi. Note: The coefficient of discharge (inside outlet opening) for a smooth and rounded outlet = 0.9, square and sharp outlet = 0.8, and square and projecting into barrel outlet = 0.7. The coefficient for this example is 0.9. For both outlets flowing and a pitot pressure of 14 psi, the total flow is 2 × 698 = 1396 gal/min. Actual flow is 1396 × 0.9 = 1256

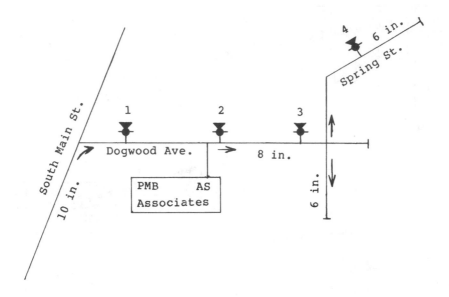

Figure 9.4   Sketch of flow test (not to scale).

gal/min at residual pressure of 42 psi. This can be graphed on semiexponential graph paper, and flows can be calculated at any residual pressure (see Figure 9.5).

While the fire protection engineer or designer should be aware of the capabilities of plotting water supply, the subject does become too involved for a general treatment on fire protection such as this. Reference can be made to the NFPA *Fire Protection Handbook* (1), FM *Loss Prevention Data Books* (2), and other hydraulics references.

An analysis of the availability of water to certain points will also indicate deficiencies that can be accounted for only by such conditions as tuberculation in piping, restricting the effective diameter of the pipe, or perhaps by other obstructions; closed valves; and occasionally the discovery of an error indicating that the pipe was actually of a smaller diameter than assumed. To find the problem areas and to ultimately provide solutions, it is recommended that a hydraulic gradient be established for the pipe in question. This gradient will provide an indication of whether a pipe may be cleaned to increase the C factor, whether a valve is actually closed, or whether it is advisable to replace a section of pipe to improve the hydraulic gradient. A *hydraulic* or *pressure gradient* is a line establishing the pres-

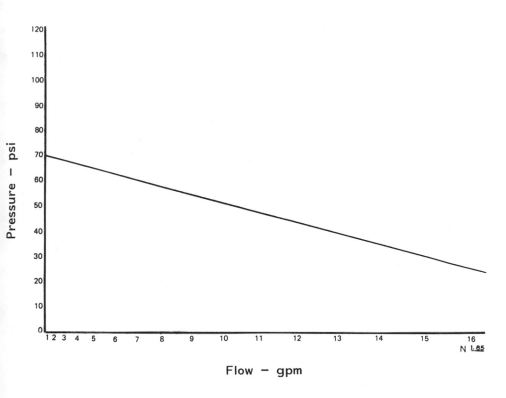

Figure 9.5 Graph of flow test. (Note: flow scale = × 100 gal/min).

sure head at any point along the pipe. The establishing of the gradient requires both fieldwork and computations.

The term *C factor* that we have referred to is widely understood in fire protection to refer to a coefficient used in the Hazen-Williams formula for determining friction loss, indicating the roughness of a pipe interior. The higher the C factor, the lower the friction loss.

WATER DEMAND

An analysis of the water supply to determine whether it is adequate will vary from plant to plant, depending upon several factors:

Construction
Fuel loading or extent of hazards

Overall values and concentration of values
Availability and accessibility of manual suppression capabilities

Therefore, the water demand must be analyzed for each plant, although we can generalize and provide a guide for water requirements as is done in the *Standard For Installation of Sprinkler Systems*, NFPA 13(5). The standard incorporates both a table or guide to water-supply requirements for pipe schedule sprinkler systems and a table and design curves for densities, areas of operations, and water-supply requirements for hydraulically designed sprinkler systems. Hose stream demand must be added to the water supply for sprinkling, as determined from the curves, to obtain the total water demand.

Occupancies are grouped in the standard in various classifications of light hazard, three groupings of ordinary hazard, extra hazard, general storage warehouses, and high-rise buildings.

The light hazard class includes those properties where the amount and the combustibility of the contents is low, such as schools, churches, hospitals, etc., where we might anticipate a fire loading of under 100,000 Btu/ft.2.

Ordinary hazard, Group 1 category includes those properties where combustibility is low, with no flammable liquids or other quick-burning materials, having stockpiles not exceeding 6 to 8 ft, and with factors generally favorable. Some examples of this classification are bakeries, automobile garages, and laundries. Also included in this category are such classifications as foundries, smelters, and steel mills. Normally, sprinkler protection is not provided in the main areas of such facilities, because of the low fuel loading and the disadvantage of the possible contact of water and molten metal, releasing hydrogen. With considerable experience and exercising good judgement, sprinkler protection may be omitted or provided only as needed in any of the occupancy groupings or portions of these facilities where there is a definite lack of combustibility from construction or the occupancy involved.

Ordinary hazard, Group 2 are those properties where combustibility of contents and ceiling heights are generally less favorable than those listed in Group 1, but with only minor amounts of flammable liquids and essentially no obstruction. Under this category are textile, knitting, and weaving mills; shoe factories; and various metal-working operations, although these may be placed in Group 1 category if they evidence favorable conditions.

Ordinary hazard, Group 3 includes those properties where combustibility of contents, ceiling heights, and construction are unfavorable; examples of this category include feed and flour mills, paper mills, piers, and wharves.

The extra hazard category includes those occupancies where a hazard is severe; this category would include extra hazard chemical operations, aircraft hangars, etc.

The value of permitting leeway to designers in laying out systems and providing systems to better meet the hazard or hazards they are designed to protect against has been recognized in NFPA 13(5). These systems, termed *hydraulically designed* or *calculated* systems, are discussed in Chapter 10, Extinguishing Systems and Equipment.

In the earlier example of a flow test (Fig. 9.4), assuming the plant is a light-hazard occupancy, the required flow (based on NFPA 13) would be 750 gal/min at the base of the sprinkler riser and 15 psi residual pressure at the highest sprinkler (5). Assuming the plant ceiling is 20 ft above the ground level, the residual pressure required at the base of the riser would be 15 psi + (20 psi/2.31 ft of water per psi) = 24 psi. The required water supply would be 750 gal/min at 24 psi residual. From the graph (Fig. 9.5), we can see that the water supply can deliver 1600 gal/min at 24 psi, which is more than adequate for the sprinkler system.

There will be occasions when the analysis of the public water supply discloses that it is inadequate with relation to the flows and pressures available. If this situation is encountered, the first approach should be to investigate with the municipality the possibility of improvements to the municipal water supply. In many cases, the municipal authorities will be fully cognizant of deficiencies in their system, and plans may be in process for improvement, or the economic impact of the proposed facility might be such that the municipality will see the desirability of providing water improvement.

Quite often, however, the inadequacy of the municipal water system will indicate the need of a private water supply, which will take the form of a fire pump taking suction from a ground tank, reservoir, or pond; a gravity tank; or possibly a booster pump connected in a bypass of a city connection (assuming that the city connection has adequate volume, but poor pressure). Again, these items will be discussed in more detail in Chapter 10.

Insurers involved in providing coverage for the HPR market frequently require a secondary water supply, with this requirement based upon the values that they are insuring at very low rates. This extra protection is required to afford reliability in the event of an impairment to one of the water supplies. As outlined above, the secondary water supply will generally consist of an on-site pump and tank, possibly a vertical turbine pump drafting from a deep well or from a pond, but may be from multiple connections to a municipal system if it can be shown that an impairment at one point in the municipal system will not cause a loss of water to the plant. In other words, if it can be determined that there is no single failure point that could affect the entire supply to the facility, two or more city connections will furnish the redundancy in water supplies required for these large facilities.

Two reliable water supplies may be necessary, depending on the basis of value and the importance of the property; the area, height, and construction of the building; the occupancy; the strength and reliability of the primary supply; and outside exposures.

REFERENCES

1.   Cote, Arthur E., ed., *Fire Protection Handbook*, 16th Edition, National Fire Protection Association, Qunicy, Mass., 1986.
2.   *Loss Prevention Data Books*, Factory Mutual Engineering Corporation, Norwood, Mass., 1989.
3.   *Simplified Water Supply Testing for Fire Departments and Insurance Engineers*, American Mutual Insurance Alliance, Chicago, Ill., 1964.
4.   *Fire Flow Tests*, American Insurance Association, New York, 1963.
5.   *National Fire Codes*, National Fire Protection Association, Quincy, Mass., 1989.

10

Extinguishing Systems and Equipment

In Chapter 2, Characteristics and Behavior of Fire, we determined from an examination of the fire triangle and tetrahedron that fires can be extinguished by any one of several methods:

Lowering the temperature below the ignition point, which requires a cooling agent
Exclusion of oxygen or air, which requires a blanketing or smothering effect
Removal of fuel, such as shutting off the gas supply
Breaking or inhibiting the combustion chain or reaction

Before a decision can be made on which extinguishing method will be used for protection, an analysis of the hazards encountered in a building should be made. This analysis will indicate the proper agent to use for extinguishment and the performance requirements to properly effect extinguishment or control. In the case of flammable liquid hazards, agents suitable for Class B hazards will be necessary. With high-piled storage, hydraulically designed in-rack sprinklers may be required.

AUTOMATIC SPRINKLER SYSTEMS

Any discussion of extinguishing systems must start with the automatic sprinkler. Automatic sprinkler protection should be considered basic and should be installed in virtually all situations. "Automatic sprinklers are the major safeguard against large losses. The evidence is substantial: during the five-year period 1981–1985, facilities that lacked sprinklers suffered an average fire dollar loss that was four times greater than those facilities with adequate sprinkler protection." (1).

In certain situations, the principal hazard may be one that is not readily extinguished or controlled by the discharge of water from sprinkler heads. For example, in an area containing basically flammable liquid hazards, automatic sprinkler protection is still highly desirable for protection of the building elements and for cooling of structural steel, although some other means of extinguishing may be used, such as carbon dioxide or dry chemical, to handle the specific hazard.

An *Automatic sprinkler system* may be defined as a system of piping of various sizes to which sprinkler heads are attached and that is connected to a water supply such as a municipal water main, gravity tank, pressure tank, etc. The sprinkler heads are activated by the thermal effects of a fire and, once open, permit the discharge of water over the fire area (See Fig. 10.1). The sprinkler system is the

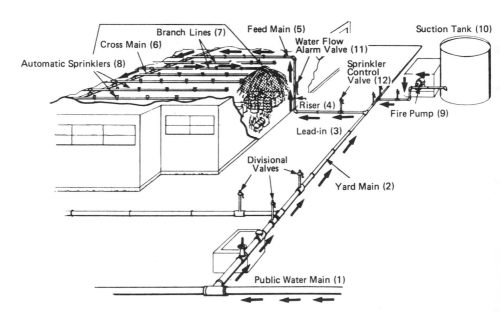

Figure 10.1 Typical automatic sprinkler system. (Reproduced with permission of Factory Mutual Engineering Corporation.)

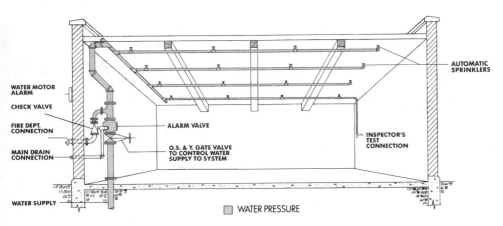

Figure 10.2 Wet-pipe automatic sprinkler system. (Courtesy of Grinnell Fire Protection Systems.)

most efficient way to deliver water to a fire, certainly exceeding the delivery of water by manual hose streams, which may be difficult to bring to bear directly on the seat of the fire. Generally, sprinkler protection stands far above other protective systems from the standpoints of efficiency and economics. It is reasonably safe to state that sprinkler protection is the foremost protective system with regard to life safety as well as property conservation.

Types of Systems

The main types of sprinkler systems are wet-pipe, dry-pipe, preaction, and deluge systems, with all but a relatively small percentage installed falling within the first two categories (wet-and dry-pipe systems). Another system, which is found very infrequently, is the combined dry-pipe and preaction system.

Wherever possible, wet pipe sprinkler systems should be provided because water is delivered faster and they are also less costly than other systems. Obviously, this type of system cannot be installed where freezing conditions may exist; consequently, dry-pipe or preaction systems must be provided in these situations.

Wet-pipe systems have water in the piping that, upon the opening of a head or heads, discharges immediately on the fire (Fig. 10.2). Dry-pipe systems, of course, have similar components and are arranged and designed in the same fashion as wet-pipe systems (Fig. 10.3). These systems differ from wet-pipe systems in that they contain either compressed air or nitrogen and, upon actuation of a sprinkler head or heads, release trapped air to the outside, tripping a dry-pipe valve to permit water to enter the series of piping and discharge eventually

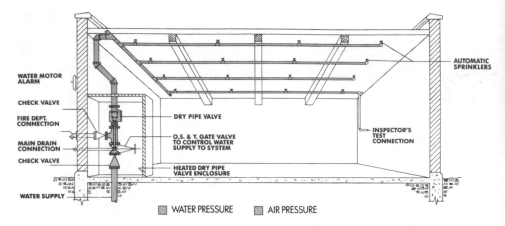

Figure 10.3 Dry-pipe automatic sprinkler system. (Courtesy of Grinnell Fire Protection Systems.)

through the fused heads. The disadvantage here is the delay in releasing water on the fire plus the cost of providing an air compressor and a more expensive dry-pipe valve in the place of an alarm check valve or merely an vane-type water flow alarm.

Dry-pipe systems can be made to operate more rapidly by providing a quick-opening device at each dry-pipe valve consisting of an accelerator or an exhauster. The devices most commonly used today will discharge air into an intermediate chamber of the dry-pipe valve, unbalancing the system and permitting the dry-pipe valve clapper to open, releasing water into the piping.

Dry-pipe valves should be used only where heat cannot be supplied. Where they must be installed in unheated warehouses to protect high-piled storage, the hydraulic design of a system will require a greater area of application due to the delay in the discharge of water. In the design of a hydraulically calculated dry-pipe system, a lower C factor of 100 is used for calculating friction loss, which results in greater pipe sizes than a wet system, which may use a C factor of 120. The term *C factor* is frequently encountered in hydraulics and is a factor indicating the roughness of interior piping in the Hazen-Williams formula most commonly used in fire protection hydraulics for determination of friction loss. A high C factor of 140 or 150 is used for smooth interiors such as cement-lined or plastic pipe, whereas cast iron pipe in service many years and tuberculated pipe could have a C factor of 50 or 60.

The preaction system is used where more sophisticated equipment such as computers may be located and where the possibility of an accidental discharge of water from a sprinkler head cannot be tolerated. This type of system employs the usual arrangement of piping with closed sprinkler heads, but also requires a detection system that, when alarmed, permits a preaction valve to open and allow water to flow into the system. No water, however, will be discharged until a head is fused as in a conventional dry-or wet-pipe system. Thus, an accidental dislodging of a sprinkler head will not cause water to flow out of the system because the detection system has not alarmed and, therefore, has not opened the water control valve. Because the head or heads may be dislodged from preaction systems without anyone's knowledge, it is usual to provide a low air pressure within the piping for supervision of the system. These systems, particularly in view of the detection required, are considerably more costly.

Up to this point, all the sprinkler systems that we have described use closed-head sprinklers. Deluge system on the other hand, use open sprinklers or nozzles attached to a piping system with a water supply that is controlled by a valve actuated by the operation of a detection or heat-responsive system installed in the same areas containing the open sprinkler heads. It can be readily seen that when the valve opens, water will flow through the piping and discharge through all of the sprinkler heads simultaneously. These systems are used mainly for special hazards such as flammable liquid exposures in the petrochemical field and other chemical operations, the protection of large-capacity oil filled transformers, and the protection of aircraft hangars. More recently, they have been used for high-piled rack storage where the pallet load has been shrink-or stretch-wrapped with plastic film.

The *Standard for the Installation of Sprinklers*, NFPA 13 (2) for the installation of sprinkler systems should be considered the basic source of information for anyone involved with the design, installation, inspection, or maintenance of sprinkler systems. In addition, the *Recommended Practice for the Inspection, Testing and Maintenance of Sprinkler Systems*, NFPA 13A (2), should be consulted on the care and maintenance of such systems. *Automatic Sprinkler Systems*, 29 CFR 1910.159, (3), covers the requirements for sprinklers when such systems are required by a particular OSHA Standard. Compliance with NFPA 13 is considered as acceptable compliance with the OSHA regulation.

Prior to 1972, NFPA 13 (2) was principally a standard calling for the installation of automatic sprinklers on the basis of pipe schedules laid out for light hazards, ordinary hazards, and extra hazard occupancies. Installations based on these systems have compiled an excellent protection record when connected to adequate water supplies, but in many cases were inefficient or were not performance oriented. In 1972 and 1973, the NFPA Committee on Automatic Sprinklers made sweeping changes that included emphasizing the hydraulic design of sprinkler systems. While this emphasis now exists, it should be noted that pipe schedules still remain in the standard principally to permit installation of small systems without requiring that they be calculated. It is anticipated that these changes will ultimately result in virtually all sprinkler systems being hydraulically designed, which should result in better designs probably at lower cost than systems installed in accordance with a particular piping schedule.

Hydraulically Designed Sprinkler Systems

A hydraulically designed sprinkler system is defined in NFPA 13 as "one in which pipe sizes are selected on a pressure-loss basis to provide a prescribed density (gallons per minute, per square foot) distributed with a reasonable degree of uniformity over a specified area" (2).

These systems are designed to deliver a minimum density in gallons per minute per square foot over a specified area of application. Since this area of application is at the most remote point in the system, an average density actually will be greater than the design density of the system.

The selection of a design density and an area of application requires consideration so that the proposed system will adequately extinguish and/or control the particular hazard. For high-piled storage, whether of a palletized or rack configuration, these designed densities and areas of application are fairly well spelled out in the *Standard for Indoor General Storage*, NFPA 231, and the *Standard for Rack Storage of Materials*, NFPA 231C (2), previously mentioned in rack storage. NFPA 13 furnishes a guide for determining density and areas of sprinkler operation for other occupancy classifications that were discussed in Chapter 9.

An NFPA 13 table specifies minimum gallons per minute per sprinkler plus minimum inside and outside hose stream demands for the various occupancy classes and the duration of the water supplies and minutes. The significance of the reduction in water supplies is exemplified by the Ordinary Group I category, which requires a total water supply of only 4200 gal, which is readily obtainable even from a relatively small pressure tank. In addition to NFPA 13, Factory Mutual *Loss Prevention Data Books* (4) provide detailed information on design criteria for sprinkler systems for various types of occupancies.

From the design criteria in the form of a density in gallons per minute per square foot over a specified area of application and a hose stream demand also in gallons per minute, the approximate water demand required to meet the design requirements of the system can be calculated. Matching this water demand against the available water will determine whether it is economically practical to design a system on the basis of the existing water supply. For example, assume that our design criterion is 0.30 gal/min/ft^2 for the most remote area of application of 4000 ft^2 with a hose stream demand of 500 gal/min. This will indicate that a total water demand in the area of 1700 gal/min is required, which includes 1200 gal/min as the supply required for sprinkler operation and 500 gal/min for hose streams. Obviously, if we have a water supply available at the site of, say, 800 gal/min at 30 psi residual pressure with a static pressure of 80 psi, it is impossible to provide this design on the basis of the available water.

The amount of water for hose streams provided in a hydraulic design is a judgement that must be based upon the specific hazard involved. Such a judgement will include an analysis of the fire loading, the rate of heat release, and other factors. For very light fire loadings well under 10 lb/ft^2, only a small-diameter hose may be sufficient and a hose stream demand in the area of 100 gal/min will be adequate. Moving up the scale to a moderate occupancy category where one might anticipate fire loading in the area of 20 lb/ft^2, hose stream demands in the range of 250 to 500 gal/min are advisable. Higher fire loadings require from 500 to 750 gal/min and, in some cases with very severe loadings, 1000 gal/min for the hose stream demand may be necessary.

The matter of hydraulic design calculations of sprinkler systems is beyond the scope of this chapter, but it is suggested that those interested in developing some expertise in this area consult publications of sprinkler manufacturers and the Factory Mutual *Loss Prevention Data Books* (4). Today, complicated hydraulic designs are not calculated manually but with the use of various computer programs.

NFPA 13 (2) covers requirements for high-rise buildings with automatic sprinkler system design for life safety and fire protection in buildings of noncombustible or fire-resistive construction. Even though high-rise buildings defined as those in which fire must be fought internally because of height are specifically covered, the design criteria can be applied to similar occupancies, all other things being equal. For light hazard occupancies, a maximum protection area per sprinkler of 225 ft^2, which fits a 15 by 15 ft modular concept is permitted. For light hazard occupancies, the design density is 0.10 gal/min/ft^2, and for compartmented ordinary hazard occupancies, 0.15 gal/min/ft^2. The area of application for hydraulically designed systems in high-rise buildings need not exceed 1500 ft^2, except that areas of application in ordinary hazard compartmented occupancies should be based upon a calculated area of application equal to the area of the largest ordinary hazard compartment, but not less than 1500 ft^2 nor more than 3000 ft^2. Designers are obviously taking full advantage of these reductions, which assume that the fire will not extend beyond the compartmented area. Application of this protection will result in very low water supplies being required when one considers that the minimum demand period specified is 30 min. Assuming for a light hazard occupancy the design criteria of 0.10 gal/min/ft^2 over an area of application of 1500 ft^2 excluding hose stream demand, the total water supply required will be in the area of 4500 gal (150 gal/min × 30), obviously very easily obtained as indicated heretofore by a variety of means including pressure tanks. Also, a common water supply for both standpipes and sprinkler systems requiring a minimum demand of 500 gal/min is permitted in NFPA 13 (2).

Sprinkler Piping

For the most part, steel piping is used in sprinkler systems, but copper tube and plastic pipe are being used particularly in light hazard occupancies (e.g., office buildings, schools, dwellings). The various piping components are shown in Figure 10.4. They include the riser, which is a vertical supply pipe in a sprinkler system; the feed main, which is the main shown supplying the cross mains; the cross mains, which are those pipes supplying the lines in which the sprinklers are located; and the branch lines, which are those piping lines extending from the point of attachment of the cross main to the end sprinkler, in which sprinklers are directly placed.

While in the previous paragraphs the discussion has centered on the advantages of hydraulically calculated systems, it is important to keep in mind the arrangements of feeding sprinkler piping that are provided in installations of any type. These feed systems are shown in Figure 10.5. The center central feed type has the advantage of providing more uniform discharge throughout the system. The side central feed should be the next choice if center central feed system cannot be installed. The central end feed is in the next lowest priority category, and finally, the side end feed. Central distribution is particularly desirable where sprinkler installations include a large number of heads on branch lines. With regard to sprinklers on branch lines, the schedules require that these heads not exceed eight sprinklers on either side of a cross main, except when sprinklers are installed above and below a ceiling and for extra hazard schedule where sprinklers on branch lines are restricted to a maximum of six. The reason for restricting the number of heads on branch lines in preengineered systems is that the friction loss in small piping becomes excessive. Despite this, the trend today is to use a piping loop at the ceiling or multiple loops to supply piping containing sprinkler heads, resulting in the installation of considerable smaller piping.

Sizing of risers as well as other piping components is dependent upon the number of sprinklers to be supplied by a system. Again, this is reflected in the piping schedules, which indicate that risers, in the case of light hazard schedule systems, are not required to be greater than 4 in. with no restriction on the number of sprinklers for 4-in. risers, with the exception that the area served by any one 4-in. riser on any one floor or in any one fire section shall not exceed 52,000 ft². For the ordinary hazard schedule, one system is restricted to 52,000 ft² for an 8-in. feed, except that for solid piled storage in excess of 15 ft in height or palletized or rack storage in excess of 12 ft, the area served by any on 8-in. pipe or tube size should not exceed 40,000 ft². This restriction applies for both steel or copper pipe. The schedule for ordinary hazard occupancies should be consulted for different size feeds and restrictions. The schedule for extra hazard occupancies requires that no system fed by an 8-in. pipe or tube on any one floor of one fire section exceed

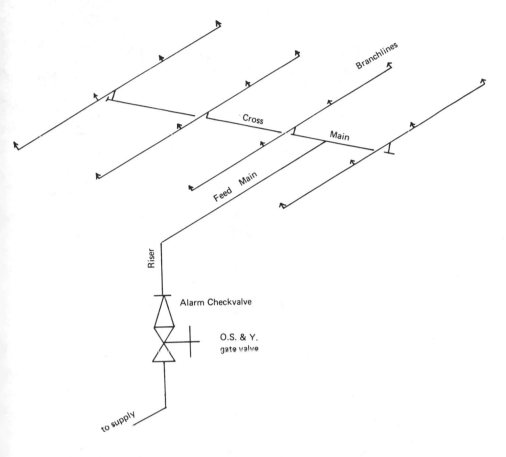

Figure 10.4 Piping designations or components comprising a sprinkler system.

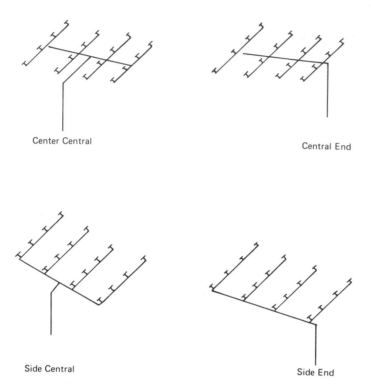

Center Central

Central End

Side Central

Side End

Figure 10.5 Sprinkler system conventional feed arrangements.

25,000 ft^2 in area. There is a corresponding reduction in the maximum number of sprinklers permitted on other sized feeds.

Risers should be located on the side of the building, and feed mains to cross mains should be run overhead from the risers. Piping installed beneath the concrete ground floor should be avoided. In the event of leakage, water flow has been known to undermine foundations and to cause extensive damage. However, if it is absolutely necessary to run piping underground, good valve control should be provided so that in the event of leaks, these areas can be isolated without impairment of the balance of the protection.

Risers should preferably be equipped with approved alarm check valves that provide for positive alarm service and a 2-in. drain for draining the system (Fig. 10.6). Alarm service is obtained through an orifice in the valve seat of the alarm check valve that permits water to enter upon lifting of the seat due to an imbalance

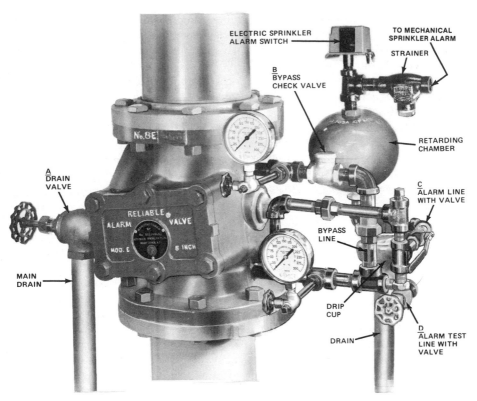

Figure 10.6 Alarm check valve on wet-pipe sprinkler system with typical trimmings. (Courtesy of Reliable Automatic Sprinkler Co.)

of pressure, with water conveyed through a retard chamber that is provided to prevent alarms due to slight variations of pressure. Alarm is then given through a hydraulic-type gong by the rotating of an alarm clapper, caused by the water pressure (called a *water motor gong*) or by the actuation of a pressure-type electric switch. Vane-or paddle-type water flow detectors are frequently installed and do not require the installation of an alarm check valve. These alarms are subject to damage and impairment by sudden surges of water such as would take place in dry pipe, preaction, or deluge systems. Surges of water are also found where pumps are provided without the use of jockey or makeup pumps. Distribution systems subject to variations in pressure will cause false alarms unless the alarm is retarded. In all these instances, it is undesirable to have a vane-type alarm installation.

It is preferable to have an alarm indicating water flow locally and also at some constantly attended location. A constantly attended location is normally an approved central station, although there are many areas in this country not covered by approved central station service, and in these cases, proprietary alarm systems or remote or auxiliary alarm service to local fire or police departments are often provided. Figure 10.7 shows the various types of supervisory services, devices, and water-flow alarms.

All valves controlling water supplies to sprinkler systems should be of an approved indicating type. These valves are located on either the exterior or interior of buildings; however, from the standpoint of being able to shut off a sprinkler system that may involve ruptured piping due to explosions, etc., it is recommended that control valves be located outside, 40 ft from the building, to prevent a problem of access to the valves. Valves located outside that control sprinkler systems, as well as controlling sections of a plant underground, should be post indicator type. Valves located in a building are of an outside screw and yoke (OS & Y) or butterfly type. Valves should be supervised by one of four methods:

1. Central station, proprietary, or remote station alarm service (See Fig. 10.7)
2. Local alarm service that will cause the sounding of an audible signal at a constantly attended location
3. Locking valves open
4. Sealing of valves and approved weekly recorded inspections when valves are located within fenced enclosures under the control of the owner

While the lock arrangement (item 3) does not specify the type to be used, one major HPR insurer requires the use of case-hardened locks. Breakable shackle locks are also available, but are infrequently used today, being much more prevalent a number of years ago. The sealing of valves in an open position, whether done with plastic or wire straps, will indicate whether a valve has been operated. Major insurers leave different colored seals at the plant, so that with the resealing of a valve in an open position after a valve has operated, it can still be determined whether the valve had been operated between inspections.

Types of Sprinklers

At the heart of any sprinkler system is the sprinkler head. In 1952, heads were redesigned and the "spray" head, so called at that time, became the accepted head providing improved water distribution. These heads, through a redesign of the deflectors, improve the spray pattern to roughly that of a half-sphere completely filled with water spray. With this type of head, little or no water is discharged upward to wet the ceiling, as shown in Figure 10.8. This sprinkler head has now

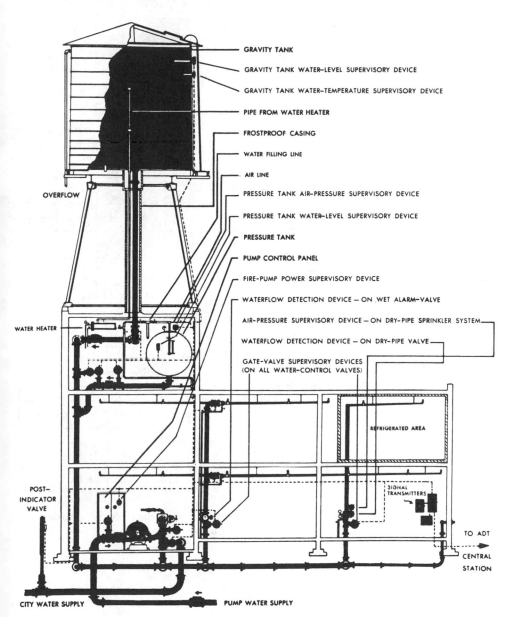

Figure 10.7 Supervisory services and devices and water-flow alarms. (Courtesy of ADT Security Systems.)

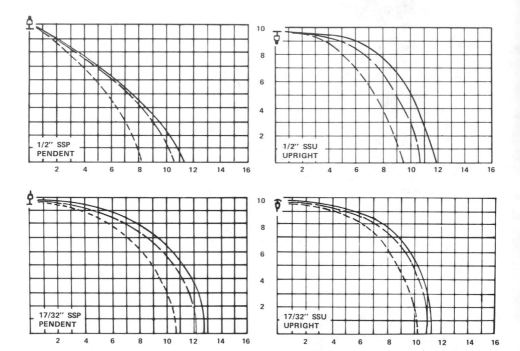

Figure 10.8 Typical water distribution pattern from a standard upright sprinkler head. (Courtesy of Reliable Automatic Sprinkler Co.)

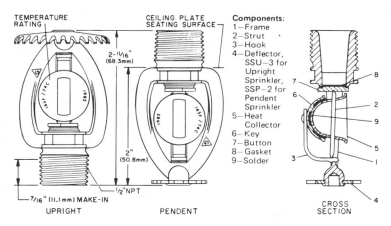

Figure 10.9 Automatic sprinklers, upright and pendent deflector designs. (Courtesy of Grinnell Fire Protection Systems.)

become known as the standard sprinkler and the sprinklers manufactured prior to 1952 as old-style sprinklers. These heads provide, at a distance of 4 ft below the deflector, a discharge covering a circular area of useful intensity of water discharge of approximately 16-ft diameter when discharging at 15 gal/min. The area covered is larger at distances over 4 ft and smaller at distances obviously less than 4 ft. Sprinkler heads that are installed in the upright position bear the indication on the sprinkler deflector of standard sprinkler upright (SSU). Sprinklers designed for use in a pendent position (deflector in a downward position) are designated as standard sprinklers pendent (SSP) (Fig. 10.9). The standard sprinkler orifice is 1/2 in.; however, oversize sprinklers of 17/32 in. are used in many hydraulically designed systems because they have a 140% greater discharge.

Automatic sprinklers come in various temperature ratings, with ordinary temperature heads used where ceiling temperatures are not anticipated to exceed 100° F (38° C). With some noninsulated metal buildings, particularly in southern climates, these temperatures can be exceeded; so-called intermediate heads that are colored white are then provided, which are installed where maximum ceiling temperatures of 150° F (66° C) are anticipated. High, extra high, very extra high, and ultra high temperature heads are also available for maximum ceiling temperatures of 225 to 475° F (107 to 246° C). High temperature heads of 286° F (141° C) are also desirable for use in ceiling systems for rack storage up to 25 ft in height. The reason for this exception is to reduce the number of peripheral heads that may be opened by a flue fire. Similarly, higher rated temperature heads may also be advisable where a hazard might be expected to produce an unusually rapid rate of heat release, such as in a flammable liquid occupancy.

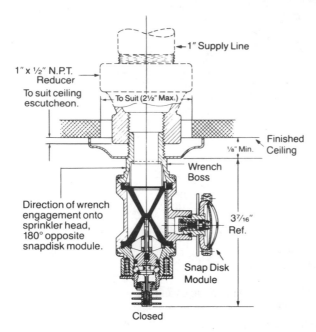

Figure 10.10   Component parts of the on-off sprinkler head. (Central Sprinkler Corporation.)

Where sprinklers are installed below open-grated mezzanines or where there are more than one layer of in-rack sprinklers, in the case of rack storage, sprinklers should be provided with water shields to prevent the sprinklers from being wetted by discharge from sprinklers located above, thereby preventing their operation.

A minimum of 18-in. clearance should be maintained between the top of any storage and sprinkler deflectors, except that 36-in. clearance is required between sprinkler deflectors and the top of storage in excess of 15 ft high in solid piles or 12 ft high in rack or palletized storage, except where the sprinkler system is hydraulically designed, where 18 in. again is permissible.

Sprinklers manufactured by various firms have different engineering principles, with the most prevalent type using a low melting point alloy. Some of these heads are faster acting than others.

Cycling sprinklers are exceedingly fast in operation and have the ability to restrict or cut off the flow of water upon extinguishment of the fire. This sprinkler head has a snap disk thermostat that actuates a pilot cylinder, permitting flow upon reaching of the prescribed temperature rating of the head (Fig. 10.10). Upon cooling, the snap disk thermostat shuts off the flow of water. The head has particular adaptations in data-processing areas and other areas where high-valued equipment susceptible to water damage is involved and also in inaccessible areas where frequent fires might be anticipated that would require the frequent replacement of heads.

Another type of sprinkler, the large drop sprinkler, has a greater discharge and produces large water drops that can penetrate strong thermal updrafts of high-challenge fires. Sprinklers have been developed for residential use that have a low-mass fusible element that results in a greater speed of operation than conventional sprinklers. Some styles operate in one-fifth of the time. Using this fast response capability, Factory Mutual Research Corporation developed the early suppression-fast response (ESFR) sprinkler. This sprinkler is designed to quickly discharge a large amount of water onto the fire. Unlike standard and large drop sprinklers, which control the fire, this sprinkler is designed to suppress the fire. This can result in less water needed and less fire and water damage. These sprinklers were originally designed for warehouse applications to eliminate the need for in-rack sprinklers. New applications for the ESFR sprinkler are being explored, both for other fire challenges and for less severe applications in manufacturing and mercantile occupancies (5, 6).

NFPA 13 (2) should be consulted for spacing and protection limitation areas for sprinkler heads. With regard to protection area limitations, for light hazard occupancy and hydraulically designed sprinkler systems, the protection area limit for a sprinkler area is a maximum of 225 ft², as previously indicated. A protection area of 200 ft² per sprinkler head is permitted for sprinklers under a combustible suspended ceiling, provided that a full complement of sprinklers is located in the space immediately above such a ceiling and the space is unfloored and unoccupied. For open wood-joisted construction, the protection area per sprinkler head for this hazard category is a maximum of 130 ft². For all other types of construction, the protection area is a maximum of 168 ft². The protection area limitation for ordinary hazard occupancies is 130 ft², except of high-piled storage, where the protection area per sprinkler should not exceed 100 ft². For extra hazard occupancy, the protection area limitation is 90 ft² for any type of building construction, except that protection area per sprinkler head should not exceed 100 ft² when the system is hydraulically designed. Certain highly hazardous occupancy classes such as storage of plastics may require a much reduced area limitation to give adequate protection and a more economical design. This will be particularly true in the case of very heavy water density requirements, such as 0.60 gal/min/ft².

With regard to the position of sprinkler heads, wherever possible, sprinklers should be used in an upright position, although frequently the suspended ceiling pendent type is necessary. With dry-pipe systems, sprinklers should be installed osition. Where pendent positions required, a special dry pendent installed.

it has been indicated that it is undesirable to provide dry-pipe sys-raulically designed system because they are slower acting and, thererore, require a greater area of application. Dry-pipe systems are limited in total number of sprinklers and in volume by NFPA 13 (2), which requires that not more than 600 sprinklers should be controlled by one dry-pipe valve and also that one dry-pipe valve should control not more than 750-gal system capacity. The exception to the volume limitation is that a system may exceed 750 gal if water is delivered to the inspector's test pipe in not more than 60 sec starting at the normal air pressure in the system. A second exception is that a system may exceed 750 gal where check valves are installed in branches of the system to assist in more rapidly reducing the air pressure above the valve seat to the dry-pipe valve trip point. With such an arrangement, no system branch should exceed a 600-gal capacity.

No dry-pipe system should be set alternately wet and dry because corrosion will accelerate in this situation. Similarly, as a result of condensate in hydraulically calculated dry-pipe sprinkler systems and anticipated corrosion in excess of a wet system, high friction losses can be expected. The Hazen and Williams formula is used when calculating friction loss in dry-pipe systems using a C factor of 100 as opposed to a C factor of 120 for black steel pipe in other than dry-pipe systems and 140 for copper tube and cement-lined cast iron pipe. This will usually result in the necessity of sizing pipe somewhat larger, resulting in higher costs. It may be possible to gain permission from the authority having jurisdiction to use a C factor equal to 120 where nitrogen is used in place of compressed air in dry-pipe systems, which will remove the problem of condensate and corrosion.

Water Supply for Sprinkler Systems

The determination of water requirements (water demand) for various hazards is largely a matter of judgment, which should be exercised only by those experienced in this area. The total water demand, as discussed in the previous chapter, includes the water required for discharge from automatic sprinklers, the water discharged from interior hose streams if they are available, and the water discharged form exterior hose streams; that is, hose streams used by either a trained fire brigade or a fire department.

In many occupancies, this total water demand will be beyond the capabilities of a municipal system; consequently, a private water supply is required that usually takes the form of a fire pump and suction tank. Gravity tanks have the limitation of furnishing relatively low pressures, although they are highly reliable. Where occupancies require fire flows for long durations, such as 3 or 4 hr, the cost of a gravity tank is usually prohibitive. Because of these low pressures, economical designs of hydraulically calculated sprinkler systems using gravity tanks are unlikely. Pressure tanks have the obvious limitation of very small volume or capacity, although they furnish high pressure.

A municipal connection is often the primary and sole supply for sprinkler systems, except for the ability of a fire department to manually pump water into a system through a fire department hose connection. Some municipalities require that fire flow be metered or that indications be provided of fire flow; this will require the use of full flow meters or detector checks. This hardware should be approved, because it is designed so that friction loss is considerably lower than the loss encountered in nonapproved equipment.

Fire Department Hose Connections

The matter of fire department hose connections is of considerable importance for sprinkler systems. Generally, for small systems, they will be located on exterior walls of buildings. However, it is preferable that connections of this type be located away from buildings to permit access by fire departments under fire conditions. A fire department connection will have a straightway check valve; an automatic means for removing water downstream of the check valve, such as a ball drip; and the actual hose connections, which, of course, should match the threads used by the local fire department (See Fig. 10.3). This hose connection for a small system will discharge into the riser, downstream from the control valve and alarm check, and should be properly labeled.

In larger facilities, multiple municipal connections are frequently made, particularly when a facility fronts on two or more streets. At each one of these connections, it is desirable to provide a pit for the fire department connections. A typical pit arrangement will be fitted with valves on either side of a check valve, which is needed to prevent water being pumped into the private system from circulating back into the city system. The pit is preferred because check valves should be serviced every five years. A pit installation will readily permit this work by merely closing the valves. The installation of the fire department hose connections in the connection feeding several sprinkler systems, generally through an underground system, is, of course, preferable in that one fire department hose connection will supply any of the systems that are operating, thereby reducing the number of fire department hose connections required. Confusion at the time of a fire is also avoided because the fire department does not have to decide which connection to hook up to.

Back-Flow Preventers

Occasionally, a pond, reservoir, or other nonpotable water supply will be advisable at the site in quantities suitable as a water supply. This supply may be acceptable as a primary supply with a diesel-driven fire pump, preferably with the pump having a flooded suction (a supply developing a head pressure at the pump suction). A word of caution is in order at this point for those who may be contemplating the cross connection of potable and nonpotable supplies. Whereas a relatively few years ago many states would accept the cross connecting of potable and nonpotable supplies with a double check-valve arrangement in the connection from the potable supply, most health authorities are prohibiting cross connecting or are requiring the use of back-flow preventer. These valves operate on a reduced pressure principle; a typical valve is shown in Figure 10.11. When a normal flow condition exists, both check valves in the back-flow preventer are open, and the pressure differential relief valve is closed. When flow ceases, pressure in the zone between the check valves is maintained at a lower pressure than the inlet pressure. When the inlet pressure drops to this lower point or less, the pressure differential relief valve opens to the atmosphere. When the outlet pressure exceeds the inlet pressure, both check valves and the pressure differential relief valve are tightly closed so that no back-flow can occur. To maintain its own pressure at the lower pressure established (lower than the inlet pressure), the pressure differential relief valve opens momentarily if the inlet pressure drops below the differential. After the required differential is established, the pressure differential relief valve again closes. Approved back-flow preventers are listed by UL in its *Fire Protection Equipment Directory* (7) that is issued annually. Listed types provide lower head friction loss than a nonapproved types. A problem with some of these approved units is that they must be removed from the line for maintenance, which may not be the case with the nonapproved type. This may indicate a desirability of provid-

Figure 10.11   A typical reduced-pressure back-flow preventer. (Courtesy of Cla-Val Co.)

ing the nonapproved type if the head loss through the unit is within acceptable limits.

Municipal Systems

The most obvious and least costly water supply for sprinkler systems is a connection to a municipal system. As pointed out, to obtain an indication of the capabilities of the municipal system, water tests must be taken to see if the system is capable of meeting the demand of the automatic sprinkler protection plus hose streams that may be required. The sizing of the municipal connection will depend upon the number of heads supplied by the system; the hydraulics in the event that the municipal connection is feeding a hydraulically calculated system; and local water company requirements, which possibly may stipulate that a connection to a municipal main be of a size smaller than the actual main size. A 4-in. connection is usually acceptable only for a very small system or light hazard systems.

Private Water Systems

Where a municipal service is inadequate to supply the required demand, the installation of a private water system must be considered. There are three methods of obtaining pressures in a system:

1. Gravity tanks or reservoirs at higher elevations
2. Pressure tanks
3. Fire pumps and tanks

Gravity and Suction Tanks   Up until the 1950s, it was common to think in terms of a gravity tank when installing a private water supply (See Fig. 10.7). Today, however, when high volumes of water at strong pressures are often needed, as in the case of high-piled rack storage or other special hazards, these tanks are no longer feasible from the standpoint of supplying adequate pressure or sufficient quantities of water such as a 3-hr supply. Economic considerations will indicate the desirability of some other means of furnishing water. In this case, it is advisable to turn to the installation of a fire pump, generally taking suction from a ground-level suction tank (See Fig. 10.1).

The capacity of tanks used for fire protection is calculated on the basis of gallons available above the outlet opening. The net capacity between the outlet opening and the overflow pipe should be at least equal to the rated capacity. Although tanks of various sizes are fabricated rather frequently, standard sizes of steel tanks range from 5000 to 500,000 gal net capacity. In calculating the required capacity of a tank, the refill capacity, if the tank is provided with automatic refill such as from an altitude valve-controlled city fill may be deducted. For example, assume a water demand of 2000 gal/min for 3 hr with the tank having an automatic refill capability of 500 gal/min. Neglecting the refill rate, a total water requirement of 360,000 gal would be calculated, resulting in the need for a

350,000 or 400,000 gal tank, as opposed to calculating on the basis of deducting the refill capability from the total demand, giving a total water requirement of a 270,000 gal or a 300,000 gal tank.

It can be seen from the preceding example that if we wish to use gravity tanks to supply our total capacity, cost would, in all probability, preclude the installation of a 400,000 gal gravity tank on a trestle that would give sufficient pressure to meet our demand. This would probably mandate the use of a ground suction tank. Design information on both gravity and ground suction tanks may be found in the *Standard For Water Tanks for Private Fire Protection*, NFPA 22 (2), and also in the Factory Mutual *Loss Prevention Data Books* (4).

Generally, the fire protection engineer will not get involved in the actual structural design of tanks, although certain items are worthy of mention from the standpoint of providing some greater basic knowledge for those not involved to the extent of the fire protection engineer in such problems or installations. The foundations for pump suction tanks must be given some consideration; these tanks are set on crushed stone, sand, or concrete foundations. Under good soil conditions, a minimum 6 in. of crushed stone or sand is provided and laid on moistened and compacted gravel after removing the topsoil portion. A concrete ring wall is constructed around the perimeter of the tank beneath the tank shelf extending below the frost line and at least 2.5 ft in depth and 6 in. in thickness. The ring wall should project 4 in. above grade. Where poor soil conditions are encountered, an 8-in. reinforced concrete slab is required under the tank bottom and within the ring wall. Under extreme conditions, piling may be required to support the tank.

Steel gravity tanks are supported by towers or by large risers (steel plate pipes 3 ft or more in diameter), as is the case with water spheres. Infrequently, it may be required to provide fireproofing of the gravity tank supports, depending upon exposure to a rather hazardous condition. The degree of fire resistance required will obviously depend upon the exposure from the standpoint of the anticipated heat release and the distance of the tank from this particular exposure. The location of a water tank, whether gravity or ground suction, should be considered from the standpoint of its exposure. Where conditions permit, the tank should be well removed from any potential danger source. Also, in the case of a multisupply system, the tank or tanks should be remote from the city connection or from each other to provide greater flexibility from the standpoint of impairments to a portion of the plant underground.

Embankment-supported rubberized fabric tanks, or otherwise ESRF tanks, are available in 20,000-gal to 1,000,000-gal capacities. This type of tank uses a reservoir liner and a flexible roof and is supported by an earth embankment on all four sides. The liner material is a nylon fabric coated with an elastomer to provide abrasion and weather resistance based on federal and ASTM design standards. Site preparation is very critical with this type of tank. After excavating for the tank, the underlayment consists of a 6-in. layer of sand or clean soil over a 3-in. layer of pea gravel (2, 6).

All tanks, whether elevated or on the ground, should have some means of determining the depth of water. These gauges require occasional checking to determine accuracy.

Both gravity and steel suction tanks require tank heating equipment unless suction tanks or elevated tanks with 3 ft or larger risers are located where the lowest 1-day mean temperature is 5° F (–15° C) or above. Data are available from the U.S. Weather Bureau for lowest 1-day mean temperatures indicating where the 5° F (–15° C) isothermal line exists and consequently, where some form of tank heating is required.

Tanks should be heated to a minimum of 42° F (6° C), and low-temperature alarms should be supervised by either an approved central station or a proprietary alarm panel (See Fig. 10.7).

In determining the size of a tank heater, it is necessary to calculate the heat loss from the tank based upon the 1-day minimum mean temperatures obtained from the isothermal map and depending upon the tank capacity and square footage of tank surface. This table may be found in NFPA 22 (2) and is based upon maintaining minimum water temperatures of 42°F (6°C) at a wind velocity of 12 mi/hr.

Pressure Tanks   In our discussions of sprinkler systems and water supplies, we determined that changes in NFPA 13 (2) in recent years have lessened water demand requirements for various hazards. These changes conceivably could encourage the use of a greater number of small-capacity pressure tanks in the future. In the case of a water-supply requirement of a total of 6000 gal, a pressure tank can be provided economically having a capacity to meet the total demand. As the name implies, these tanks are under air pressure and accordingly are fabricated under the terms of Section VIII, "Rules for Construction of Unfired Pressure Vessels," of the American Society of Mechanical Engineers (ASME) *Boiler and Pressure Vessel Code* (8). These tanks will normally contain two-thirds water with the balance air pressure under a minimum of 75 psi. The pressure required in the tank must be sufficient to give not less that 15 psi pressure at the highest sprinkler as the last of the water leaves the tank. Pressure tanks should be provided with automatic means of maintaining the required air pressure and should also be provided with alarms indicating low air pressure and low water level. The following formula should be used in determining the pressure in tanks:

When tank is placed above the highest sprinkler, $P = \dfrac{30}{A} - 15$

When tank is below level of the highest sprinkler, $P = \dfrac{30}{A} - 15 + 0.433\dfrac{H}{A}$

where

P = air pressure carried in pressure tank
A = proportion of air in tank, normally expressed as 1/3
H = height of highest sprinkler above tank bottom

In the event of the interconnection of a pressure tank and a gravity tank, air lock must be avoided. This is accomplished by connecting the discharge pipes from the tanks 45 ft or more below the bottom of the gravity tank.

Fire Pumps   To this point, we have discussed two of the means of developing pressures to gravity and pressure tanks—and we not turn our attention to the third means, the use of pumps. Two types of centrifugal pumps are used in fire protection systems, horizontal split-case and end-suction types and vertical shaft turbine, with the horizontal split-case centrifugal pump being by far the most common. These pumps are available at different capacities ranging from 25 to 5000 gal/min, with rated pressure heads of 40 psi or larger. Approved fire pumps are specially designed for reliability and ruggedness and, in the case of the horizontal split-case type, provide ready access to all parts. In addition, all working parts are of a noncorrodible material.

Horizontal Centrifugal Fire Pump   Horizontal split-case centrifugal fire pumps should preferably take suction under pressure or what is also commonly called *flooded suction* (see example of configuration in Fig. 10.7). These type pumps are called *positive suction*) fire pumps and can be supplied with rated net pressure of 40 psi or more. Figure 10.12 shows a typical horizontal split-case fire pump with driver. Although not desirable, horizontal split-case pumps are occasionally used to take suction under lift. These pumps are called *negative suction* fire pumps, but can also be used with flooded suction and are supplied at rated heads of 100 psi or more. Horizontal split-case fire pumps can take suction under lift only up to 15 ft, which means that the lift and friction loss in the connecting piping must not exceed 15 ft. Pumps taking suction under lift also require priming to displace the air in the pump casing and suction pipe within 3 min. It is preferable to provide a priming method that will keep the pump primed constantly. This can be accomplished by means of an automatically filled priming tank, which is probably the most common method used. The *Standard for the Installation of*

Figure 10.12 A typical horizontal split-case centrifugal fire pump. (Courtesy of Aurora Pump Co.)

Centrifugal Fire Pumps (2), should be consulted for full information not only on priming methods but also on the whole subject of centrifugal fire pumps.

Vertical Shaft Turbine Fire Pump Where water must be taken under lift from an open body of water, such as a pond or river, a vertical shaft turbine fire pump should be used. This pump does not require priming because the impellers or pump bowls are submerged at all times. The vertical shaft turbine pumps, an example of which is shown if Figure 10.13, are available in the same capacities as the horizontal split-case type and with rated pressures of 100 psi or greater.

Pump Characteristics In addition to the special design of reliability and ruggedness, fire pumps are also designed to discharge 150% of rated capacity at 65% of rated pressure. To illustrate, in the case of a pump rated at 1000 gal/min at 100 psi, the pump also has the capability to discharge 1500 gal/min at 65 psi pressure. Horizontal split-case pumps have a shutoff or "churn" pressure not exceeding 120% of rated head. Vertical shaft turbine pumps meet the same design points as do the horizontal split-case pumps, except the shutoff or churn pressure must not exceed 140% of the rated head. Pump manufacturers furnish a characteristic

Figure 10.13   Typical vertical-shaft turbine fire pump. (Courtesy of Aurora Pump Co.)

curve with the pump that shows a plot of these points indicating that it meets the criteria established for an approved fire pump. The same plot also showing a curve for the brake horsepower at the various flows and a pump efficiency curve. An example of a pump characteristic curve supplied by a manufacturer is shown in Figure 10.14.

Another category of centrifugal pumps beside those mentioned is used occasionally where water supplies are limited or the water demand may be rather small. These pumps, termed *limited service pumps,* are available in capacities of 200, 300, 400 gal/min at various pressures. Limited service pumps are limited to a capability of providing 130% of rated capacity at "overload" instead of the 150% previously indicated for the other type pumps. Maximum power required for these pumps should not exceed the limitations of a 30-hp electric motor.

Pump Location A water supply must be highly reliable to ensure its availability at the time water is needed. This reliability, from the standpoint of design, should take into consideration the safe location of the fire pump with respect to exposure from nearby hazards; a determination of the method of driving the

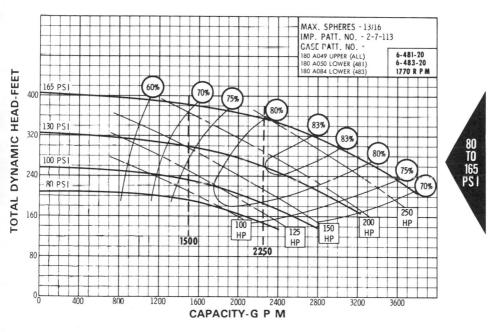

Figure 10.14 A typical fire pump characteristic curve of a 1500-gal/min fire pump with electric motor drive. (Courtesy of Aurora Pump Co.)

pump; and, in the case of an electric drive, the supplying of power to the pump in a method that will give maximum assurance against outage of the electricity, primarily due to the poor location of the electrical connection and routing of the circuit.

The location of the fire pump should not require much discussion, since it is easy to see that a combustible pump-house construction can endanger safety unless it is sprinklered. Poor placement of the pump with respect to a nearby hazardous processing area, such as a highly hazardous chemical process, and also endanger safety. In a larger plant where the pump is to be located in a detached pump house, this pump house should be of noncombustible construction and not exposed by a hazard that could directly expose the pump to outage. In the case of booster or fire pumps located within the confines of a building, the pump should be separated from combustible construction or hazards by construction having a fire rating commensurate with the adjacent hazard; that is, the combustible or fire loading exposing the pump. In most cases, an enclosure of the fire pump having a fire resistance rating of 2 hr will suffice.

Pump Drivers   Determining the method by which a fire pump is to be driven requires an analysis of the electrical power source. This analysis will indicate the degree of reliability of the electrical power source as evidenced by the past history of outages. When it has been determined that the power supply will provide the reliability desired, the next step is to ensure that the power source will be available to the pump under emergency conditions. This can be done by one of the following methods:

1. Supplying power to the pump prior to the plant load. This will ensure that disconnecting plant power will not cause a loss of power to the pump.
2. Ensuring that the electrical feed to the pump will not be exposed to damage from fire due to exposure to combustible construction or hazards or to accidental damages.
3. Providing wiring in the pump house in conduit or electrical metallic tubing.

Additional information on ensuring the integrity of an electrical supply for a pump may be found in NFPA 20 (2).

It is necessary to think about protecting the electric motors used to drive pumps from water splashing. Noncombustible shields are installed where open motors are used. Drip-proof motors will not require a shield if hose valves are located outside the pump room. Splash-proof motors will usually not require shields, and totally enclosed fan-cooled motors are also installed without this protection. All current-carrying parts of electric motors should be a minimum of 12 in. above the floor.

Internal combustion engine drives, and specifically diesel engines, while involving higher costs, should be installed where electrical installations would provide questionable reliability or where second or third pumps are installed for private water supplies to provide greater reliability or water-supply redundancy. Diesel engines are particularly reliable and should be used preferably where internal combustion engine drive is contemplated.

Pump rooms for diesel engines should be kept at a minimum temperature of 70° F (21° C) (compared to 40° F—(4° C—for electric drive). It is also desirable to employ the use of block heaters in cold areas to maintain liquid-cooled engines at operating temperatures to ensure ease in starting.

Pump Controls   Automatic control of fire pumps is desirable in most installations requiring the use of approved automatic controllers. Normally, automatic operation is caused by a pressure drop, but in situations where low-level flows will not cause any significant drop in pressure, automatic start by actuation of fire protection equipment such as sprinkler water flow is preferable.

Where automatic control is provided for pumps, the following alarm service to a constantly attended location should be installed:

Power failure
Pump failure to start
Pump running

The preceding alarms will indicate the need for attention to the pump, which can be crucial at the time of an emergency but which also can prevent emptying of tanks and burning out of pump drives in the event of undetected leaks (See Fig. 10.7).

Relief valves should be installed where adjustable speed drivers are used or where constant speed drivers are used and the shutoff pressure plus the static pressure exceeds the pressure for which the system is designed. These valves are located between the pump and pump discharge check valve and are sized according to the pump capacity.

Hose valves are normally used and are desirable for pump testing; they are connected to the pump between the discharge check valve and the discharge gate valve. The number of hose valves required will depend upon the capacity of the pump; generally speaking, on hose connection is required for each 250 gal/min of pumping capacity.

Makeup or jockey pumps of small capacity should be installed with automatically controlled pumps to avoid recycling of the fire pump as a result of leaks in the plant underground mains. These pumps are located in a bypass and usually provide 90 to 100 psi or greater pressures on systems, with fire pump cut-in pressures somewhat lower. Jockey pumps can supply only small quantities of water, and the fusing of a very limited number of sprinkler heads will cause the fire pump to change to a running mode.

Booster fire pumps, used to increase pressures of municipal supplies, should be located in valved and checked bypasses. This arrangement will permit continuity of water to a system in the event of impairment to the pump. A typical schematic of such an installation is shown in Figure 10.15.

SPECIAL PROTECTION SYSTEMS

For most facilities, sprinklers normally provide the primary defense against fire, but in certain situations, special protection systems are necessary. These special protection systems include foam, carbon dioxide, Halon, dry chemical, water spray, and explosion suppression. Basically, these are not a substitute for sprin-

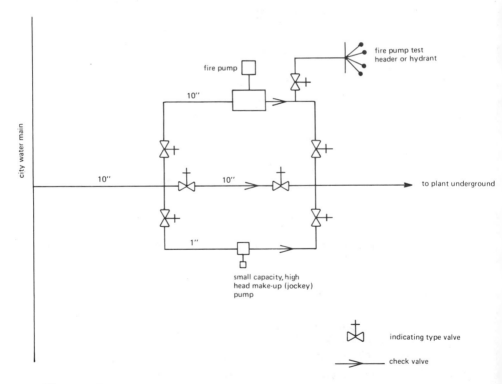

Figure 10.15 Schematic arrangement of a booster pump in a municipal water supply connection.

klers, but are there to extinguish a fire before a number of sprinklers operate. Even though these special protection systems actuate before sprinklers do, they are normally used in conjunction with sprinklers since they have a limited amount of extinguishing agent and generally protect specific hazards. Automatic sprinklers provide the long-term fire protection capability if the other systems do not extinguish the fire.

Foam Systems

The early foams were a mixture of alum and bicarbonate of soda solutions with licorice root extract and required large volumes of solution and lead-lined vessels. Next, two-powder chemical foams were developed, which were then combined into a single powder. While effective for combating fires in flammable and combustible liquids, these foams presented difficult logistical problems. Mechanically generated foams were developed in Germany in the 1930s; concentrates could be introduced into water solution more readily and also gave greater quantities of foam.

In general, fire-fighting foams extinguish by excluding oxygen from the surface, by generating a large number of small bubbles that are of lower specific gravity than oil or water. This results in the bubbles floating on the liquid surface. Fire-fighting foams can be made with different qualities to ensure a flow across the surface of a liquid and the formation of a tough or tenacious seal. The foam must have this tenaciousness to cling to both vertical and horizontal surfaces. Foams, in addition to tenacity, must have the ability to retain considerable water. In fires the water is converted to steam by the heat from the fire. The foam, in addition to the ability to flow freely and seal, must resist disruption from factors such as wind, heat, or flame and, when punctured, must be able to reseal.

Foams can be defined by their expansion ratio, which is as follows:

Low expansion—up to 20:1
Medium expansion—20 to 200:1
High expansion—200 to 1000:1

Foams, of course, have limitations. They are ineffective for protection of Class A hazards and should not be used for extremely volatile liquids, for electrical equipment unless deenergized, for liquids heated above 212° F (100° C) (which will cause frothing), for combustible metal hazards, and for materials having their own oxygen supply. Foams are basically used for combating fires in flammable or combustible liquids and, in fact, are the only practical means of extinguishing fires in large tanks containing these materials, having been known to extinguish fires in tanks having diameters as large as 140 ft.

A chemical foam has become obsolete and may be found in old systems, but is not generally recommended in modern systems. Chemical foam is produced by a chemical reaction of an alkaline salt (sodium bicarbonate) with an acid salt solution (generally, aluminum sulfate), thus producing carbon dioxide in the presence of a foaming agent, a licorice material. Carbon dioxide is trapped in the bubbles formed, producing the aggregation of small bubbles of lower specific gravity than oil or water, that is needed to seal off the oxygen from the fuel. Chemical foams come in stored solutions where the alkaline salt and the acid salt are sorted separately in solution until foam is needed, then the two materials are brought together, creating the foam in water. These systems may be small or very large and may be automatic or manual.

A number of types of foaming agents are available. These foams are produced by a foam liquid added to water to form a solution that is capable of foaming when aspirated. The foam liquid is called the *foam concentrate*; in the case of regular protein foams, it consists of hydrolized animal or vegetable protein, stabilizers, solvents, and an industrial germicide. Regular protein foams are suitable for ordinary hydrocarbon liquids and are available for proportioning in water in 3% or 6% concentrations. Protein foams provide good resistance to burn-back, good extinguishment, and relatively low cost. In addition, they are biodegradable, nontoxic, and noncorrosive and present no major problem in cleanup. They can be stored and used at temperatures ranging from 20 to 120° F (–7 to 49° C).

There are recent improvements to protein foams known as *fluoroprotein* foams, which, while more expensive, have improved compatibility with dry-chemical agents and have been used for subsurface application in tanks, since the bubbles formed form this concentrate have the quality of shedding the fuel. Fluoroproteins, which are protein with a synthetic fluorinated surfactant, are also available in 3% and 6% solutions and can be substituted in existing protein foam systems. Fluoroproteins have shown superiority over regular protein foams in fire experience while retaining the good qualities of the regular protein type.

Certain flammable liquids can deteriorate foams. The types of solvents that can destroy regular foams are called *polar* or *alcohol-type* solvents and are miscible with water or other constituents of the foams. Solvents of this category are alcohols, esters, ethers, and ketones. Special foams are required for these water-soluble liquids; they are termed polar solvent type, water-soluble type, or alcohol type foams. Generally, it may be said that polar-type foams are required if the fuel contains oxygen, nitrogen, or sulfur.

The most common types are polymeric alcohol-resistant aqueous film-forming foam (AFFF) concentrates, which can be applied at a distance above the liquid surface. This is particularly important in using foams for protecting tankers. Transit time (the time of delivery of the foam solution, containing concentrate and water, to the point of foam making or aspiration) is limited with the protein-base polar solvent types, but is not a factor with the polymer-base polar solvent type.

AFFF agents are fluorinated surface-acting agents that have very fast knockdown capabilities and may be particularly advantageous when used in fuel spills. They have been found to be 25% to 30% more effective than regular protein foams. AFFFs do have a disadvantage of being fast draining and, as a consequence, do not present long-range fuel security in comparison with other foams. They do have a very high degree of compatibility with the use of dry chemical and, in this case, exceed fluoroprotein foams.

Of considerable interest in the fact that AFFFs can be used by proportioning into conventional closed-head sprinkler systems. They actually produce a better foam from ordinary sprinkler heads than from the foam sprinkler heads that are needed for aspiration in the use of protein foams in foam or foam-water sprinkler systems. This would suggest that where flammable liquid hazards are protected by ordinary hazard sprinkler systems, an AFFF system providing storage and proportioning into the system can be used for fast knockdown and consequently considerably greater protection. Foam sprinkler systems require a density of 0.16 gal/min/ft^2 to extinguish, which most ordinary sprinkler systems can deliver. AFFFs can probably do an effective extinguishing job even at densities somewhat lower than 0.16 gal/min/ft^2, possibly in the area of 0.10 gal/min/ft^2. If there is a flammable liquid hazard being protected by an ordinary hazard system, protection can be greatly upgraded by the use of AFFFs.

Film-forming fluoroprotein agents (FFFP) combine the film-forming fluorinated surface action of AFFF with the fuel-shedding capabilities of fluoroprotein foam. They have fast spreading and leveling characteristics.

Foams are applied on exterior storage tanks through fixed foam discharge outlets, which are devices for introducing the foam into the tank. This application may be Type I, which is an approved outlet that will conduct and deliver foam gently to the surface without submergence of foam or agitation of the surface. In our discussions of foams, we noted that, in the case of protein-base polar solvent-type foams, Type I application was required. Type II application merely indicates that the foam does not need to be delivered gently but must be designed to lessen submergence and agitation. Subsurface delivery is also becoming a popular method of delivering foam to large tanks.

Fixed installations of foam systems are recommended for exterior tanks of large capacity used for storage of flammable liquids. These installations consist of a central foam house from which foam is delivered to tanks at fixed delivery outlets or to chambers attached to the tanks. This may include the use of fixed pumps. Semifixed installations are considered to be those where discharge outlets attached to the tank are connected to piping that terminates at a safe distance from the tank, where foam making apparatus may be connected.

In foam application on large storage tanks containing hydrocarbons, a minimum of 0.10 gal/min/ft^2 of foam solution (foam concentrate and water) should be supplied. Certain polar solvents require higher extinguishing rates and the *Standard for Low Expansion Foam and Combined Agent Systems*, NFPA 11 (2), should be consulted.

With the use of protein foams, the quantity of foam concentrate required may be determined by multiplying the application rate in gallons per minute per square foot by the square footage of the surface of the liquid by 0.03 or 0.06 (depending upon the concentration) times the minimum discharge time. Minimum discharge times may also be found in NFPA 11 (2), broken down into hydrocarbons and water-soluble solvents with either Type I or Type II application. Note that discharge times will be less in the case of hydrocarbons for Type I application. In the case of gasoline, naphtha, or other flammable liquids having flash points of less than 100°F (38° C) this will be a discharge time of 30 min for Type I application and 55 min for Type II application.

Refer also NFPA 11 for information on supplemental foam hose streams, which are required in addition to tank foam installation for protection of ground fires that may develop. The minimum number of hose streams and minimum operating times, based upon minimum solution rates of 50 gal/min, are also listed in a table provided in NFPA 11.

Tanks of different diameters require a different number of foam discharge outlets depending upon size. Tanks of up to 80 ft in diameter require a minimum of one discharge outlet and tanks over 180 to 200 ft in diameter require a minimum number of six discharge outlets. A varying number of discharge outlets is required for in-between diameters. To prevent loss of fuel, the discharge outlet should be provided with frangible seals designed to burst at 25 psi.

Foam systems are frequently used for protection of flammable liquid hazards located indoors, using either overhead or floor-type outlets. Spray-type foam sprinklers available for overhead applications are located on the basis of one for each 100 ft^2 of floor area. These overhead nozzles should be located to envelop the equipment to provide insulation. These systems apply the foam solution at a minimum rate of 0.16 gal/min/ft^2 in the case of rooms with a minimum discharge time of 10 min.

Low-velocity open discharge outlets may be used particularly for installation of foam at the floor level for spill protection. These open discharge outlets, which may be nothing more than open pipes, are also suitable for discharging foam on the surface of flammable liquids contained in tanks such as dip tanks. These systems may be automatic or manual. If automatic, they must be provided with a detection system for actuation of automatic valves, which should be as close as possible to the hazard. Any single system, due to water requirements, should also be kept as small as practical. As indicated, with area protection, the solution rate should be a minimum of 0.16 gal/min/ft^2, with higher applications for polar-type solvents with operating times of a minimum of 10 min.

The *Standard for the Installation of Deluge Foam-Water Sprinkler Systems and Foam-Water Spray Systems*, NFPA 16 (2), should be consulted for information on foam-water and spray systems. In these systems, the application of foam is considered primary and water discharge after the depletion of the foam is a secondary consideration providing cooling. Foam-water sprinklers or foam-water spray nozzles are used with the nozzles differing only in the design of the deflectors and providing a somewhat different pattern of discharge. The foam-water sprinkler heads provide patterns close to the discharge of standard sprinklers that make foam by aspirating the foam-water solution at the nozzle. These sprinklers are provided with minimum 1/4-in. diameter orifices. The foam-water spray nozzles supply a more directional of pattern.

In these systems, the components consist of the piping with the foam-water or foam-water spray heads, a water control valve, and a system for proportioning the foam concentrate into the water and forming a foam solution. The positive-pressure injection method is recommended for injecting the concentrate into the water. This portion of the system includes a storage tank for the concentrate, a foam concentrate pump, and a metering orifice into the riser with a "telltale" indicator listing the diameter and flow direction. Obviously, the pressure of the foam concentrate entering the system must exceed that of the system itself. For reliability, the concentrate pump is invariably a centrifugal type.

The foams we have discussed so far are always made near the point of application, with the foam solution aspirated with air to produce foam. The means of proportioning the concentrate into the water stream to provide a foam solution is normally by use of pump proportioners and Venturi induction.

No discussion of fire-fighting foams would be complete without including *medium-*and *high expansion foams* defined as an aggregation of bubbles from mechanical expansion of a foam solution by air or other gases. Foams with expansions from 20:1 to 1000:1 can be generated. Heretofore, the discussions pertaining to foam have covered only low-expansion foams, having expansion ratios of under 20:1. Medium-or high-expansion foams are used as a flooding agent mainly for enclosed volumes, being specifically of interest in protecting hazards that have shielded spaces or voids where conventional protection agents would have difficulty reaching a fire. These foam systems provide a three-dimensional approach to fire fighting by forcing water into these difficult-to-reach areas. While medium-or high-expansion foams are used for exterior fires, they are used principally for combating indoor fires and, in the case of Class A materials, provide control when the hazard is completely covered with foam. Extinguishment will be accomplished if the foam is sufficiently wet and is maintained for a sufficient soaking period. In the case of Class B flammable or combustible liquid fires, high-flash-point liquids can be extinguished when the surface is cooled below the flash point of the material, with lower flash point materials extinguished when the foam blanket is of sufficient depth over the liquid to exclude oxygen.

Medium-or high-expansion foams extinguish by blanketing when provided insufficient volume excluding air; by reducing the oxygen content when the water of the foam is converted to steam; by a cooling effect from the conversion of water to steam, which absorbs heat from the fuel; and through the penetrating of Class A materials. Further, the foam has the quality of insulating materials not involved in the fire. Medium-or high-expansion foam may be used for flammable and combustible liquid hazards or for high-piled storage of such materials as rubber tires or rolled paper where considerable voids are present that are inaccessible to the usual extinguishing agents.

This type of foam is generated by spraying a concentrate on a screen and then forcing large volumes of air through the screen, forming bubbles. The bubbles formed may be passed through ducts or chutes or dumped directly into the enclosure.

The detergent-type concentrate should be listed for use with the equipment being utilized and is generally used at temperatures ranging from 35 to 110° F (2 to 43° C). The listing of the generator will also indicate the concentrate required to be used with the equipment, the water required in gallons per minute, and the discharge rate of the generator in cubic feet per minute. Minimum water pressures at the generator are also supplied in the listings, which are generally in the vicinity of 50 psi.

While the generator should be placed as close as possible to the hazard, special protection of the generator may be advisable. These generators may be mounted on outside walls, on rooftops, or in adjacent rooms. Ducts for airflow can be furnished to the atmosphere if location requires. Medium-or high-expansion foam generators are either electric motor driven or water driven, with the larger capacity generators being driven by an electric motor.

Medium-or high-expansion foam systems will consist of a means of detecting a fire, a means of actuating foam generators, a means of closing doors and opening of vents near the roof line to remove any displaced vapor in the air, and the means of sounding necessary alarms and supervisory signals.

The *Standard for Medium and High Expansion Foam Systems*, NFPA 11A (2), specifies certain foam requirements for protection of interior hazards, including a minimum depth of foam above the hazard that is 1.1 times the height of the highest hazard, or a minimum of 2 ft above this hazard.

The time to reach submergence height of volume is also specified in NFPA 11A, depending upon the hazard involved, the nature of the construction, and whether the building is provided with sprinkler protection. Sprinkler protection is desirable, although sprinklers will destroy some of the foam. This is offset by discharge of water providing protection until the submergence volume is reached and also protecting structural members. The rate of discharge involves a determination of the construction of the building, with lighter construction more susceptible to heat and requiring a faster buildup of the foam to prevent collapse of steel.

The rate of discharge or total generator capacity for high-expansion form is determined from the following formula:

$$R = (\frac{V}{T} + R_S) \, C_N \, C_L$$

where

R = rate of discharge, ft³/min
V = submergence volume, ft³
T = submergence time, min
R_S = rate of foam breakdown by sprinklers, ft³/min
C_N = compensation for normal foam shrinkage, usually a factor of 1.15
C_L = compensation for leakage of foam through openings in the structure

The rate of foam shrinkage due to the discharge from sprinklers is determined by multiplying the foam breakdown in cubic feet per minute per gallon per minute of sprinkler discharge given in the listing of the equipment (usually 10 ft³/min/gal/min) times the estimated total discharge from sprinklers. In the case of a hydraulically calculated sprinkler system, this discharge will be the density multiplied by the area of application.

The quantity of concentrate needed to generate high expansion foam system should be sufficient to provide either 25 min of generation—i.e., to provide for generation for four times the submergence volume (basically, volume of hazard to be protected) or a minimum of 15 min of operation. The submergence volume should be at least 60 min for nonsprinklered locations and 30 min for sprinklered locations. These submergence times may be considerably reduced if only flammable liquids are involved.

While it is true that for relatively small enclosures, very limited quantities of water are required for a high-expansion foam extinguishing system, this is not true for very large volumes. For example, a 73,500 ft³/min foam generator unit has a water requirement of 542 gal/min. Assuming a large volume to be protected by a high-expansion foam system requiring a total generator capacity of 400,000 ft³/min, a water requirement in excess of 2300 gal/min will be necessary to supply the generators. This is often not recognized, since there appears to be a general impression that little water is required for any high-expansion foam system.

Carbon Dioxide Systems

Carbon dioxide extinguishing systems are frequently found in industry for protection of special hazards. The special hazards may take the form of flammable liquids and protection of electrical and electronic equipment located in control rooms, computer underfloor areas, complete computer envelopes, etc. Carbon dioxide extinguishes primarily by reduction of the oxygen supply with a low cooling capacity when applied directly on the burning material; consequently, for local application, it is not considered desirable for extinguishing fires in Class A materials where a wetting or extended cooling action is required. However, carbon dioxide systems are used for protection of record vaults and, in some instances, for providing protection for libraries, particularly where rare books are involved. These are exceptions and in these instances, a total flooding system is required for flooding an entire volume; this system must be tightly constructed to prevent CO_2 leakage to maintain the carbon dioxide in a desired concentration for a relatively long "soaking" period.

Carbon dioxide, as indicated, has certain limitations. The first is that it has no appreciable cooling effect and, therefore, is undesirable in combating Class A hazards. Another drawback, particularly as opposed to a sprinkler system, is that carbon dioxide systems provide only limited quantities of extinguishing agent and, consequently, must extinguish immediately unless provided with a reserve supply. Without a reserve supply, these systems are out of service until cylinders can be recharged and the system restored to operation.

Carbon dioxide cannot be used for extinguishment of fire in all materials. It is well recognized that these systems extinguish by reducing the oxygen content. Some materials, however, contain their own oxygen supply that is released under combustion conditions such as cellulose nitrate. In addition, certain metals and metal hydrides can decompose carbon dioxide. Typical examples of these reactive metals are magnesium, titanium, zirconium, and alkali metals such as sodium and potassium.

Also because it reduces oxygen content, carbon dioxide should not be used in normally occupied spaces and in spaces not normally occupied but where personnel may be present for other purposes, unless arrangements can be made to *ensure evacuation* before discharge. This is extremely important, since evacuation may be difficult due to large spaces and complicated passageways, or egress may be impeded by obstacles. This difficulty increases once the carbon dioxide discharge begins due to possible confusion caused by the noise of the discharge and the greatly reduced visibility. In addition, the carbon dioxide may leak or flow into unprotected lower levels such as cellars or pits, creating an oxygen-deficient atmosphere (6).

While it is advisable to point out the limitations of CO_2 systems, let us now direct our attention to the positive aspects of CO_2 installations and indicate why carbon dioxide is often desirable as an extinguishing agent. First of all, CO_2 is electrically nonconductive and inert. In addition, it is colorless and odorless and, even though stored in a liquefied form, it gasifies leaving no residue. Therefore, it certainly can be considered a clean agent.

Carbon dioxide systems can be categorized as a high-pressure or a low-pressure system. Low-or high-pressure systems can be applied by either total flooding or local application. High-pressure systems use liquid carbon dioxide stored in cylinders at 850 psi at 70° F (21° C), while low-pressure systems use liquid carbon dioxide stored at 300 psi at 0° F (–18° C). Normally, for smaller systems, the high-pressure types are used for reasons of economy. For larger systems, it is probably more economical to use low-pressure systems. However, these systems require larger containers that must be refrigerated.

Low-pressure systems are generally considered more economical if the volume to be flooded exceeds 2000 ft³. If space is a factor, the low-pressure storage container, which generally is of considerable size, can be a detriment, as opposed to the usual high-pressure carbon dioxide cylinders, ranging from 5 to 100 lb, which require generally little room and may be installed tightly against a wall.

Total Flooding

As indicated above, both low and high-pressure carbon dioxide systems can also be total flooding or local application types. A total flooding system consists of a supply of carbon dioxide connected to fixed piping with nozzles arranged to discharge carbon dioxide into an enclosure around the hazard. A local application system will discharge carbon dioxide directly on the hazard, such as the surface of a dip tank. To determine whether a local application or total flooding system should be used, let us refer to two examples. The first example is a single dip tank located in a large room. Obviously, it is advisable to consider local application protection from the standpoints of both economy and life safety, since the concentrations of carbon dioxide used in total flooding present a life safety hazard from anoxia or asphyxiation. In the second example, however, let us assume that we have a room where we have several dip tanks, and where we may have to change the location of these dip tanks frequently. Under this condition, a total flooding system that ejects carbon dioxide into the entire room to inert the complete volume is more practical and provides for complete flexibility in the continued relocation of the equipment.

In total flooding systems, the quantity of carbon dioxide for extinguishment varies depending upon whether the fire can be characterized as a surface-burning one or the hazard might be anticipated to develop deep-seated fires. First let us confine our discussion to surface-burning fires where, in most cases, it is sufficient to supply enough carbon dioxide to reduce the oxygen content to 15% for extinguishment. However, certain materials require greater reduction of oxygen content, some to as low as 6%. Minimum carbon dioxide concentrations for extinguishment of various flammable liquids and gases and other materials may be obtained form various publications, including the *Standard on Carbon Dioxide Extinguishing Systems*, NFPA 12 (1). This information will usually be presented in table form with two columns, one the theoretical minimum carbon dioxide concentration percent and the other column the minimum design concentration of CO_2 in percent, which is the theoretical minimum concentration plus 20%. For example, acetylene has atheoretical CO_2 concentration of 55% and a design concentration of 66%. For materials not provided in tables, the minimum theoretical carbon dioxide concentration is normally obtained from a recognized source such as the U.S. Bureau of Mines or is determined by test. If maximum residual oxygen values are available, the theoretical carbon dioxide concentration may be calculated using the following formula, with the design concentration obtained by merely adding an additional 20%:

$$\% \ CO_2 = \frac{(21 - O_2)}{21} \ 100$$

Carbon dioxide needed for surface-burning hazards requiring minimum design concentrations of 34% or less is determined by applying volume factors. These factors indicate the cubic feet per pound of CO_2 or its reciprocal, pounds of carbon dioxide per cubic foot of the enclosure.

Smaller volumes of space require greater quantities of carbon dioxide per cubic foot because these smaller volumes have proportionally more boundary area and greater leakages can be anticipated.

The enclosure of a hazard protected by a total flooding system must be reasonably tight to avoid excessive loss of the extinguishing agent. This loss must be compensated for by additional CO_2 equal to the anticipated loss during a 1-min holding time.

For deep-seated fires, it must be mentioned again that under normal conditions, water is the most suitable agent for extinguishing fires. However, as indicated previously, water can do extensive damage to valuable items such as rare books; consequently carbon dioxide is frequently used for protection of these types of enclosures were high concentrations of values might be anticipated. For deep-seated fires, it is necessary to maintain carbon dioxide for a considerably longer period of time; in fact, the enclosure should be tight enough to retain the design concentration for at least 20 min. This means that no nonclosable openings should be permitted except for small openings located at or in the ceiling. It might be well at this point to mention that openings in enclosures surrounding a hazard can be closed automatically using carbon dioxide as the closing medium by providing pneumatic trips actuated by the pressure of the gas.

Understandably, the volume factor for slow-burning materials will require greater quantities of carbon dioxide.

It is suggested that the arrangement and sizing of pipe and orifices of nozzles be left to those who are expert in the design of carbon dioxide systems. However, some comments pertaining to these installations are in order. Since the flow of carbon dioxide is not completely liquid, the pressure drop through piping is greater near the end of the pipeline than it is at the beginning. CO_2 leaves the storage container as a liquid at container pressure; as the pressure drops because of friction, the liquid boils, producing a mixture of liquid and vapor. Curves such as found in the appendix of NFPA 12 (2) are used by designers or by others for checking flow rates.

Local Application

With regard to local application systems, the nozzles for discharging carbon dioxide may be attached to the sides of dip tanks or may be located overhead to discharge carbon dioxide on the liquid surface. The carbon dioxide requirements will depend upon the area of the hazard to be protected and the arrangement of the nozzles. Briefly, after laying out the hazard, the designer will refer to the listing information for the particular nozzles to be used, which is available in the UL publication *Fire Protection Equipment Directory* (7), published annually. The listings will provide information in the case of overhead nozzles showing, at specified heights above the hazard, and flow rate in pounds per minute and the area of coverage in square footage for the same height. With regard to tank side nozzles, the listing will give the area covered per nozzle in square feet plus the flow rate in pounds per minute per nozzle in both minimum and maximum quantities. Once the nozzles are laid out the flow rates are obtained, the total flow rate is obtained by simple addition. It is then merely required to determine the discharge time for the hazard, which will be a minimum of 30 sec (but which could be longer depending upon the hazard) to determine the total quantity of carbon dioxide required to provide protection for the particular hazard. In the case of high-pressure systems, the quantity of carbon dioxide must be increased by 40% to determine the actual cylinder storage capacity because only the liquid portion of the discharge is effective. This increase in cylinder storage capacity is not necessary for total flooding systems.

It should be noted that the information on flow rates in the UL *Fire Protection Equipment Directory* (7) is also available from the manufacturers' listing or approval curves. These listings will show flow rates versus the hazard width or the flow rate versus the area coverage in the case of tank side nozzles. With overhead nozzles, the listings will show maximum area in square feet of coverage versus the height or distance from the liquid surface and also the flow rate versus the height or distance from the liquid surface.

System Components

Let us briefly examine other components of automatic carbon dioxide systems, starting with the detection devices. Obviously, these devices should be placed in locations where they will give prompt notice of heat, smoke, flame, or combustible vapor conditions. A determination of the type of detection needed will depend upon the hazard and could take the form of heat or smoke detectors. Detection equipment will be described in Chapter 11, where the advantages and disadvantages of each type will be outlined. The location is extremely important, and these detectors should be installed within the maximum spacing requirements as outlined by the UL *Fire Protection Equipment Directory* (7). Equally important is the power supply for electrical detectors from the standpoint of reliability and independence from the electrical supply to the hazard area.

The detection devices must provide for release of carbon dioxide from the storage container, which is done by releasing or actuating devices that are electrical, mechanical, or pneumatic. In the case of high-pressure cylinders, this actuation or release of the carbon dioxide is done either by pneumatic control heads or by electrical control heads, depending upon the type of detection used. Carbon dioxide stored in these cylinders is a liquid; at 70° F (21° C), the pressure is at approximately 850 psi, and the internal pressure acting on the cylinder valve holds it tightly closed. In the event of fire, an electric control head operates by the closing of the thermostat within the head; in a pneumatic control system, the temperature increase acts upon a diaphragm that trips the control head and causes it to operate. Operation of the control head causes a plunger in the control head to move forward, opening the pilot seat in the cylinder valve and releasing carbon dioxide through the pilot opening to a channel that leads to a chamber above the discharge head piston. The pressure of the CO_2 on a spring-loaded piston opens the main valve seat. Figure 10.16 shows a typical CO_2 cylinder control head.

In more cases than not, the size or type of hazard requires several cylinders of carbon dioxide in the high-pressure system to be manifolded together. In systems of this type with more than two cylinders, two of the cylinders should be fitted with control heads that are actuated by the detection devices. The other cylinders are known as *slave* cylinders, and these cylinders have pressure-operated or discharge heads that open by means of the pressure of the discharge of the carbon dioxide from the control or master cylinders.

Frequently, it is desirable to use one manifold of carbon dioxide cylinders; that is, one supply of carbon dioxide for the protection of multiple hazards. This requires the use of directional or selector valves that also may be either electrically or pneumatically operated, depending upon the detection system used. An example of a selector or directional valve is shown in Figure 10.17. The use of these valves can be a considerable cost savings, requiring only one common supply for a number of hazards. The use of directional valves and one supply requires that no

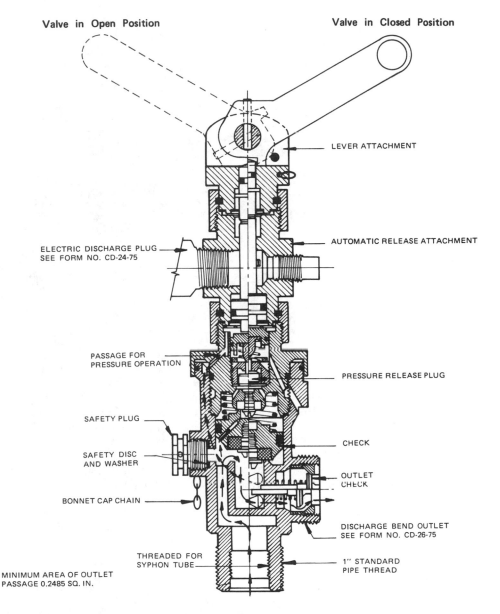

Valve in Open Position   Valve in Closed Position

LEVER ATTACHMENT

AUTOMATIC RELEASE ATTACHMENT

ELECTRIC DISCHARGE PLUG
SEE FORM NO. CD-24-75

PASSAGE FOR
PRESSURE OPERATION

PRESSURE RELEASE PLUG

SAFETY PLUG

SAFETY DISC
AND WASHER

CHECK

BONNET CAP CHAIN

OUTLET
CHECK

DISCHARGE BEND OUTLET
SEE FORM NO. CD-26-75

THREADED FOR
SYPHON TUBE

1″ STANDARD
PIPE THREAD

MINIMUM AREA OF OUTLET
PASSAGE 0.2485 SQ. IN.

Figure 10.16 Control head for carbon dioxide systems. (Courtesy of Ansul.)

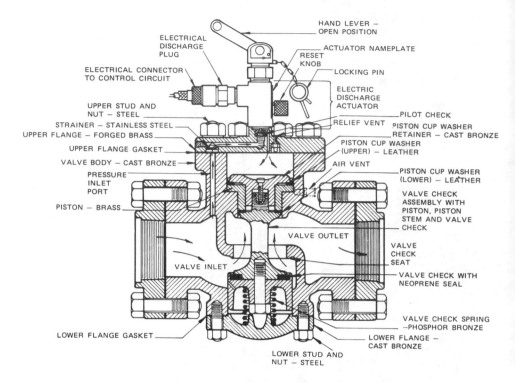

Figure 10.17   Selector valves used in a carbon dioxide protection system. (Courtesy of Ansul.) Open position: Valve is shown in the open position. The valve actuator attachment opens the pilot check, permitting gas to flow into the upper piston chamber as indicated by the arrows. The relief vent to atmosphere is closed when the valve is in the open position and the upper chamber is pressurized. After discharge: The pressure in the upper piston chamber will exhaust throught the pilot check. As the pressure decreases in the upper chamber, the spring under the valve check will close the valve. The relief vent serves to exhaust any residual pressure in the upper chamber. When discharge has been completed, the valve actuator attachment should be reset.

more than one of the hazards will be simultaneously involved in a fire. If this is not the case,the quantities of carbon dioxide provided must be sufficient for protection of both hazards.

In addition to providing for automatic actuation of the carbon dioxide systems, provisions must also be made for manual actuation. In a pneumatic control head, this is done by a cable release; in an electric control head, by a simple switch-type control.

Most insuring authorities desire the use of a reserve supply of carbon dioxide that can be placed in service upon the actuation of the main supply by merely operating a manual switch. This necessitates a check valve between the supplies. The determination of whether to provide a reserve supply will depend upon the need for continuity of the process protected, an analysis of the hazard, and the availability of refill capabilities within the area. Normally, most hazards located where there are 1-day refill capabilities do not require a reserve supply.

Some additional comment is warranted about the life safety aspects of carbon dioxide when used in total flooding systems. It is well documented that a carbon dioxide concentrations above 9%, personnel could lose consciousness. At concentrations of about 20%, death would follow in about 20 to 30 min unless the victim were removed to a source of fresh air. From our previous discussions, we determined that the minimum concentrations generally used are around 34%, considerably above the danger point. This dictates that safeguards be provided to protect personnel in areas subject to the discharge of carbon dioxide, including an unmistakable alarm and a delayed discharge. The amount of delay in releasing CO_2 depends upon the area involved and the ease of egress from the protected location. The alarm should be arranged to sound continuously; in addition, consideration should be given to the ease of escape from the area from the standpoint of adequate lighting and directional signs, adequate aisle space, the ease of opening doors exiting from the area, and detailed instructions and training of the employees on the personal action they must take when the alarm sounds. Employees should be warned of the noise of the discharges from gas extinguishing systems (including Halon systems) to help allay the fright that may be caused by the sudden roar of the discharge.

Halon Systems

Halon fire-extinguishing systems suppress fires very effectively when designed for specific fire protection uses. Halons are hydrocarbons in which one or more hydrogen atoms have been replaced by atoms from the halogen series: fluorine, chlorine, bromine, or iodine. At present, the halons used in fire suppression are Halon 1211 (bromochlorodifluoromethane), Halon 1301 (bromotrifluoromethane), and Halon 2402 (dibromotetrafluoroethane). Halon 1211 and Halon 1301 are liquified, compressed gases; Halon 2402 is a liquid. They are electrically nonconductive and have high liquid densities, which permits use of compact storage containers. They rapidly vaporize in fire and leave no corrosive or abrasive residue. The halons have been particularly effective in data-processing (computer) areas, industrial control rooms, telecommunication centers, sensitive medical diagnostic equipment, and other employee-attended occupancies housing electrical equipment (6).

Unlike other gaseous agents such as carbon dioxide, Halon 1301 does not extinguish by smothering but by inhibiting the combustion chain reaction. While this mechanism is not completely understood, Halon 1301 breaks down in flame into two radicals, trifluoromethyl and bromine. One theory suggests that the bromide radical reacts with the fuel to form hydrogen bromide, which further reacts with the free hydrogen atoms in the reaction zone to produce hydrogen molecules. The hydrogen molecules disperse or burn, forming water and bromine, and these reactions are repeated, resulting in the removal of active hydrogen atoms and depriving the fire of fuel.

Unfortunately, these same bromine atoms that fight fires so effectively also can break up ozone molecules in the atmosphere, similar to the reaction of the related chlorofluorocarbons (CFC) with ozone. Concern for the earth's diminishing ozone layer has forced government and industry to reassess the value of halons as fire suppressants.

In 1987, 24 nations, including the United States, signed the *Montreal Protocol of Substances that Deplete the Ozone Layer*, which identifies and regulates five chlorofluorocarbons and three halons: Halon 1211, 1301, and 2402. The EPA, as signatory for the United States, adopted the protocol in full as EPA regulations. The protocol recommends exploring the use of alternative fire-suppression systems before committing to halons. Where a halon system is essential to property loss control, halon discharge should be reduced through (1) limiting discharge quantity in fire suppression, (2) eliminating nonessential discharges, and (3) minimizing the potential for unnecessary discharges. The NFPA, FMRC, and others in the fire protection community have begun to take steps to follow the recommendations of the protocol.

In 1992, the EPA plans to enforce a cap on production and consumption of the halons at 1986 levels. In the meantime, the world's leading manufacturer of halon (50% of the production) has pledged to cease halon production in the year 2000. At present, researchers and manufacturers are actively involved in research to develop non-ozone-depleting halon substitutes (9, 10).

In a nondecomposed state, Halon 1301 is not considered to be toxic, being placed in the UL Group 6 category, the least toxic classification based upon animal exposures. In concentrations below 7%, Halon 1301 has little effect on humans, and exposure to Halon for 4 to 5 min is quite safe. At concentrations above 7%, there is a noticeable anesthetic effect, becoming more pronounced with increased concentrations up to about 15%. This effect produces mild dizziness and a sense of mild intoxication. At the concentrations below 10% that are normally found in extinguishing systems, the effect is mild. Most Halon 1301 systems are designed to provide extinguishing concentrations of 5% to 7%; therefore, it is permissible to allow employees in an area where Halon 1301 is discharged in these concentrations for relatively short periods of time. This can be a major factor when considering, for example, the need to maintain continuity of programming in the master control room of a television station.

On the other hand, the products of decomposition of Halon 1301 can be hazardous. In the theory mentioned above, in which Halon 1301 provides extinguishment by free radicals, a certain amount of decomposition must exits. Halon 1301 decomposes at approximately 900° F (482° C); therefore, there is decomposition at an interface between the agent and the flame front at concentrations of less than an extinguishing amount. These products of composition are hydrogen fluoride and hydrogen bromide, with small quantities of other chemicals that are generally too small to be of concern. With a properly engineered system of Halon 1301 that develops an extinguishing concentration in 10 sec or less, the problem of decomposition of Halon is felt to be relatively minor. It is reasonably safe to say that fires occurring in areas where properly engineered Halon 1301 systems have been provided have not shown any excessive decomposition of the agent.

Total Flooding

Halon 1301 is considered to be particularly desirable for total flooding protection of areas containing highly valuable and sophisticated electronic equipment or in areas that contain highly valuable records or materials, such as data-processing tapes (Fig. 10.18). In hazards where possible deep-seated fires might occur, it is necessary to provide higher extinguishing concentrations of Halon for a longer than normal period of time. This extended time, known as the soaking period, will prevent any burn-back as a result of the diminishing of the Halon 1301 to below an extinguishing concentration.

While the majority of halon systems are of a total flooding type, local application systems can also use this agent effectively, particularly for flammable liquids where low concentrations of Halon 1301 may be used, as opposed to the higher concentrations required when using carbon dioxide. The interior of specific data-processing equipment can also be protected in this fashion.

With total flooding systems, concentrations greater than 10% should not be used in normally occupied areas. Since the normal extinguishing concentration in total flooding systems is from 5% to 7%, this presents no problem. If greater concentrations are required, it might be advisable to consider carbon dioxide rather than the more expensive halon.

As with other types of agents used in total flooding systems, the enclosure should be reasonably tight and unclosable openings should be kept to a minimum. This usually requires the shutdown of ventilating systems.

Normally, solid materials that would be anticipated to have a surface fire only would be expected to be extinguished with concentrations in the range of 5% to 7%. Where deep-seated fires are anticipated and the need for a clean agent such as Halon 1301 is not that important, one might better consider the use of water. Where deep-seated fires may be encountered, it is advisable to maintain the extinguishing concentration of the agent for a greater period of time or a longer soaking period, rather than to raise the extinguishing concentrations significantly. Formulas determining the mass flooding factor (pounds of Halon 1301 per cubic foot of closed volume) and the volume flooding factor (cubic feet of enclosed volume per pound of Halon 1301) used to calculate the quantity of agent required after a determination of the design concentration may be found in the *Standard on*

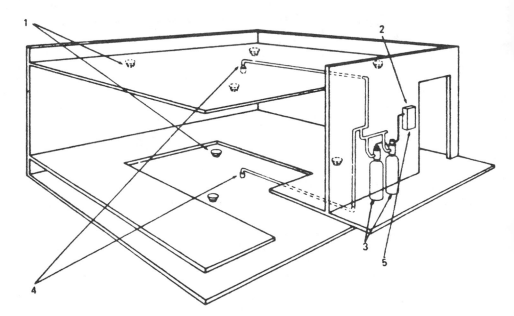

Figure 10.18   A total flooding halon system. (Courtesy of Ansul.) (1) Automatic fire detectors installed both in room proper and underfloor area, (2) Control panel connected between fire detectors and cylinder release valves, (3)Storage containers for room proper and underfloor area. (4) Discharge nozzles installed both in room proper and underfloor area. (5) Control panel may also sound alarms, close doors, and shut off power to the area.

Halon 1301 Fire Extinguishing Systems, NFPA 12A (2), which also lists these factors versus design concentrations.

Previously, we've mentioned the desirability of obtaining an extinguishing concentration as quickly as possible to prevent excessive decomposition of the agent. Design concentration should be reached in 10 sec or less; however, in most systems, it is highly desirable to reach design concentrations in 5 sec or less.

As with any total flooding system, additional quantities of Halon 1301 will be required to compensate for any loss due to the inability to halt ventilation systems or to close openings. In case the agent is lost for either of these reasons, this determination may be found in NFPA 12A (2).

The vapor pressure of Halon 1301 varies with temperatures. At a 199-psig vapor pressure at 70° F (21° C), there is sufficient pressure to discharge the halon content of storage containers. However, as the temperature decreases, the vapor pressure decreases correspondingly, so that it is necessary to pressurize the agent above the vapor pressure. This operation, called *superpressurization* is done with nitrogen to ensure adequate discharge of the agent at lower temperatures and also to ensure that the agent in the distribution system will be in a liquid stage and not drop below the vapor pressure.

Containers of Halon are superpressurized with nitrogen to either 360 or 600 psig at 70° F (21° C). Storage containers for Halon may be conventional cylinders, such as those used for carbon dioxide, containing up to 100 lb of halon, or may be high rate discharge (HRD) containers of either spherical or cylindrical configuration. High-rate discharge containers have the advantage of generally being mounted very close to the hazard and, as a consequence, require very limited piping. Containers with top-mounted valves must be fitted with an internal dip tube to expel the material in a liquid form in the same manner as liquid carbon dioxide. Figure 10.19 shows a typical HRD system-protected hazard. Noted that with these containers, the suppressant is discharged by the operation of an electrical initiator releasing a mechanical device. These cylinders, when located in a fire area, are provided with a fail safe feature created by the rupturing of a rubber gasket at elevated temperatures (approximately 300° F—149° C).

As with any gaseous automatic extinguishing system, there must be a way to detect the upset condition and actuate the release devices on the storage containers, permitting the discharge of the gas. With halon, because of the high cost of the agent, particularly when very large systems are encountered, consideration of the accidental dumping of the system is paramount. To avoid unwarranted dumping of the agent or to provide for the possibility of manual suppression prior to the actuation of the release devices, two alarm systems are often provided with the one system consisting of smoke detectors to provide a prealarm with the actuation of the system by a heat detector such as a rate compensation type. Cross-zoning of detectors may accomplish the same objective but is done primarily to prevent the inadvertent release of halon. Cross-zoning simply means that a second detector must alarm to actuate the system and requires two circuits for detectors. Alarming the first detector will give a prealarm, permitting the possibility of manual extinguishment.

Dry-Chemical Systems

Dry chemical as an extinguishing agent is very known and is applied by portable extinguishers, hand hose line systems, and fixed systems. The dry chemical used most extensively is sodium bicarbonate, which has excellent extinguishing capabilities when used to combat fires inflammable liquids, but this agent is also acceptable for use on electrical fires. The extinguishing agent does have the drawback of being corrosive and, therefore, affects finely polished metal surfaces such as found in electronic equipment.

Potassium bicarbonate (Purple K) as used in dry-chemical extinguishers or systems is a better extinguishing agent on flammable liquids than is the more conventional type using sodium bicarbonate. The same limitations exist in using the

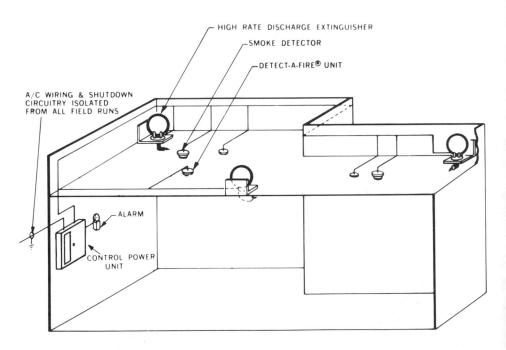

Figure 10.19 Typical Halon 1301 protection system using high-rate discharge containers. (Courtesy of Fenwal Inc.)

agent to combat electrical fires in highly sophisticated and valuable electronic equipment, suggesting the use of carbon dioxide or halon as more desirable.

In addition to the two dry chemicals mentioned above, monoammonium phosphate is used in multipurpose or "ABC"-type extinguishers. This class of dry chemical may be used on any type of fire (except Class D fires) and is particularly desirable for use by untrained personnel, because it eliminates a decision on the part of the operator as to the correct extinguisher to use on a particular fire. This material has the same disadvantage as the other dry chemicals: when it is deposited on hot metal surfaces, it forms deposits that may be difficult to remove. It also prevents the saponification of grease such as is encountered in metal cooking hoods. Therefore, monoammonium phosphate should not be used in cooking-hood applications or where highly polished metal surfaces are encountered, such as in electronic equipment.

The remaining dry chemicals in use are potassium chloride (Super-K) and urea-potassium bicarbonate.

The *Standard for Dry Chemical Extinguishing Systems*, NFPA 17 (2), specifies requirements for systems along with the use of dry chemical in hand hose lines. As with CO_2 systems, dry-chemical fire protection systems may be either total flooding or local application types. In essence, this means that in total flooding systems, the dry-chemical powder is discharged into an enclosure; with local application systems, the dry chemical is discharged directly on to the hazards, such as the surface of a flammable liquid. There are a very limited number of total flooding dry-chemical system installations, but local application dry-chemical systems are relatively common, and preengineered dry-chemical systems for the protection of cooking hoods are probably found more extensively than any other type of system (Fig. 10.20).

Because the flow of dry chemical in an expellant gas such as nitrogen does not follow the laws of hydraulics, the design of dry-chemical systems is based to a very great extent upon flow principles determined empirically. Each manufacturer produces its own dry chemical, which may not possess the same characteristics as the dry chemical manufactured by another. For this reason, the actual design of a dry-chemical system should be left to those who are competent in this field. The dry chemical used in a system should not be changed unless the change is approved by a nationally recognized testing laboratory, by the manufacturer of the equipment, or by others having authority in this area.

It has mentioned previously that dry chemical is particularly effective in extinguishing flammable liquid and gas fires, and it can be used advantageously for electrical fires (where finely polished metal surfaces are not involved), such as in oil cooled transformers and oil circuit breakers. Dry chemical is also perhaps the best extinguishing agent for use on textiles where flash fires might be anticipated. This is particularly true of potassium bicarbonate and monoammonium phos-

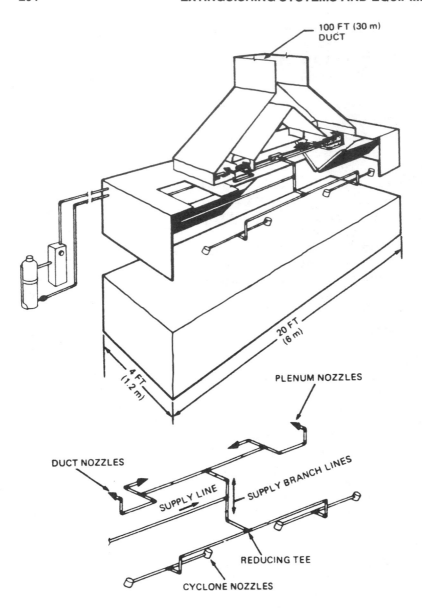

Figure 10.20 A typical preengineered dry chemical system for kitchen range, hood, duct, and fryer fire protection. (Courtesy of The Ansul Company.)

phate (multipurpose type). Multipurpose dry chemical is also effective against Class A fires of a surface-burning type and where a deep-seated fire is not anticipated to develop.

As with every extinguishing medium, dry chemical also has its limitations and is not effective on chemicals that contain their own oxygen supply, such as cellulose nitrate. Dry chemical should not be used on fires involving combustible metals such as sodium, potassium, magnesium, titanium, and zirconium.

Total Flooding

Total flooding systems of dry chemical are possible only where there is an enclosure that is reasonably tight and has limited unclosable openings not exceeding 15% of the total area of the sides, top, and bottom of the enclosure. As with other total flooding systems, because of the limited quantity of agent and its rather short-term effect, the possible sources of reignition should be eliminated. This may involve deenergizing electrical equipment. Also, openings such as doors, windows, etc., should be arranged to close with the use of pneumatic trips. Where openings having areas in excess of 1% of the total area of the sides, top, and bottom of the closure and not exceeding 5% cannot be closed, these openings must be compensated for by the use of additional dry chemical—a minimum of 0.5 lb/ft^2 of the unenclosed openings—with the additional dry chemical applied through the regular distribution system. In the event that these unclosable openings exceed 5% of the total of the sides, top, and bottom of the closures and do not exceed 15%, the additional amount of dry chemical for compensation of the loss should be not less than 1 lb/ft^2 of unenclosed openings, which is locally applied simultaneously over the openings.

As with other types of total flooding systems, ventilating systems should be stopped. Where this is impossible, additional dry chemicals should be added to the enclosure through the regular distribution system to compensate for the loss through the action of the dry-chemical exhaust.

Local Application

Many flammable or combustible liquid hazards have been effectively protected by local application dry-chemical systems, as previously indicated. The typical preengineered dry-chemical system for cooking hoods is a combination of total flooding and local application systems. In local application systems, the designer must consider the minimum quantity of dry chemical required, the minimum flow rate of dry chemical, the nozzle distribution patterns, the limitations of placement of nozzles with respect to flammable liquid surfaces, and the possible obstruction of the dry chemical distribution patterns.

Dry-chemical system design, including the layout of the piping for the system, should be left in the hands of those competent in the installation of dry-chemical systems. Special attention must be given to the arrangement of piping and changes in direction to avoid the possibility of the dry-chemical powder being thrown out of the expellant because of what is termed *two-phase flow*. This two-phase flow characteristic of gas and suspended solid depends upon the nature of the dry chemical being used, the type of expellant gas, and the design of the equipment.

Water-Spray Systems

Fixed water-spray systems are generally used for the protection of special hazards, which may include those involving flammable and combustible liquids, either in open tanks or in enclosed vessels or storage tanks, for extinguishment of Class A hazards, for protection of fire wall or floor openings from the passage of smoke and heat, for the protection of certain electrical hazards and for protection against exposure fires. Fixed water-spray systems have been also used effectively to protect rack storage of various materials to considerable heights where the product is encapsulated in a plastic film. In rack protection, water spray is applied directly into the face of the rack by "across the aisle" discharge.

From this brief description of the uses of water spray, it can be seen that this is not a system that could be considered general or basic protection, but is one that is used to combat special hazards. Fixed water-spray systems may be desirable in protecting chemical processing equipment, particularly in highly congested areas where it could conceivably be difficult to use hose streams and also where there might be an extremely rapid and high rate of energy release from a process.

Water-spray systems involve the use of detection equipment and deluge valves that must be actuated by the alarming of the detectors. The nozzles used have smaller orifices than do conventional sprinklers and, as a consequence, are more subject to possible obstruction. These systems provide for the simultaneous wetting of large areas and, therefore, should be used only where damageability of the contents from the discharge of water is within acceptable limits.

A fixed water-spray system includes the use of nozzles that are designed to provide a water-spray pattern and water droplets of a specified size, connected through a piping system to a water supply that may be either automatically or manually controlled and that in the case of an automatic system, also involves the use of detectors to actuate the water control valve.

The type of nozzles used will depend upon what is expected of the system. Different nozzles provide different patterns and different droplet size. These varied discharge patterns are furnished by wide-angle and narrow-angle discharge nozzles. Because the orifice size of water-spray nozzles is normally less than that of conventional closed-head sprinklers, usually 1/4 to 2 in. in diameter, strainers are usually provided in the supply lines to prevent clogging and individual nozzles also may have strainers incorporated in the design. In addition blowoff caps may be used to prevent depositing of foreign substances on the nozzle, which also might prevent discharge.

Water-spray systems provide an extinguishing action by several means, including cooling, smothering, emulsification of liquids, dilution of liquids, providing a layer of water over the surface of a liquid if the material has a specific gravity of greater than 1, or a combination of the above factors.

The cooling action of the water spray will depend on the the flash point of the materials involved. Liquids having flash points with temperatures below the temperature of water cannot be extinguished by cooling alone. Smothering is accomplished by the conversion of the water to steam by the heat of the fire. This conversion expands the volume by approximately 1750 times, which helps to exclude oxygen and extinguish the fire. Emulsification of liquid may be transitory, depending upon the viscosity of the materials, with a more viscous liquid providing a longer lasting emulsification effect and protection against burn-back. This action is obtained by the agitation of the impingement of the water spray on the surface of the liquids. Some dilution of water-soluble liquids will take place, but this action is usually of a relatively minor nature.

The control of fire is a legitimate design objective for certain materials that may not be susceptible to complete extinguishment. Systems so designed are usually required to provide protection for an extended period of time until the material representing the hazard is completely consumed or until the flow of a particular material may be shut off.

Exposure protection is frequently encountered using water spray in areas such as protection of liquefied petroleum gas containers.

Water-spray systems for fire prevention must be able to function for the time necessary to provide cooling, dispersing, dilution, or dissolving of a flammable liquid or other hazardous materials.

For extinguishment, the *Standard for Water Spray Fixed Systems for Fire Protection*, NFPA 15 (2), suggests a general range of water-spray application design rates that will apply to most combustible solids or flammable liquids with 0.20 gal/min/ft^2 to 0.50 gal/min/ft^2 of protected surface. This standard should be consulted for a greater in-depth treatment of this subject. The proper design density will depend upon an analysis of the material to be protected, which will include considerations such as the flash point, the degree of water solubility, the specific gravity, and the viscosity of the liquids to be protected. The standard points out that care must be exercised with very viscous materials, such as asphalt, because of the potential problem that may be created by frothing or slop-over. Additionally, care must also be taken with materials having a hazardous reaction with water, such as sodium or potassium.

For control of fires, the system should include areas where possible flammable liquid spills may travel or accumulate. These areas and other critical equipment should have water delivered at a rate not less than 0.50 gal/min/ft^2 of projected surface area.

For exposure protection, the system must be able to perform for the duration that an exposure fire may be anticipated. This requires an analysis of the exposure problem from the standpoint of the amount of combustibles and the anticipated response of manual suppression. To prevent the possible failure of the equipment from the exposure fire, NFPA 15 (2) requires that the system be designed to discharge effective water spray from all nozzles within 30 sec following the operation of the detection systems. In the case of vessels, the standard requires that water be applied to vertical or inclined vessel surfaces at a discharge rate of not less than 0.25 gal/min/ft^2 of exposed uninsulated surface, with individual nozzle water application rates required to be increased to provide for any "rundown" or slippage allowances. The densities that are specified in NFPA 15 (2) for exposure protection provide for a minimum wastage of 0.05 gal/min/ft^2.

It is quite apparent that with spherical or horizontal cylindrical surfaces, the bottom portions cannot be considered wettable from rundown from discharge above the center portions to any appreciable amount. Consequently, bottom and top surfaces of vertical vessels should completely covered by direct water spray at an average rate of not less than 0.25 gal/min/ft^2 of exposed uninsulated surface.

It is most important to protect the supports for vessels containing flammable liquids (to prevent toppling). This should be accomplished by using a discharge of not less than 0.10 gal/min/ft^2. This discharge for the uninsulated skirts or supports may be on one exposed uninsulated side, either inside or outside. For protection of structures against exposure fire, horizontal structural steel members require a minimum density of 0.10 gal/min/ft^2 over the wetted area, and vertical structural steel members require a minimum density of 0.25 gal/min/ft^2 over the wetted area, with nozzles spaced not farther apart than 10 ft on the centers. For exposure protection of pipe, tubing, cable runs, etc., a minimum rate of 0.10 gal/min/ft^2 of the aggregate pipe wall area is required.

Water-spray systems are used extensively for protection of oil-cooled transformers. Systems should be designed to discharge on all exterior surfaces at a water rate of not less than 0.25 gal/min/ft^2 of protected area of a rectangular prism envelope around a transformer and its appurtenances, with a minimum discharge rate of 0.15 gal/min/ft^2 on the nonabsorbing ground surface area of exposure. For transformers and other outdoor hazards, greater droplet size is used to overcome the effect of wind.

With water being discharged through all nozzles, the water demand for water-spray systems is generally high. The water supply must be capable of supplying the demand or design rate for the time period of discharge for all systems designed to operate simultaneously. NFPA 15 (2) imposes a limitation on the size of a single system of 3000 gal/min and indicates that separate fires areas should be protected by separate systems. These fire areas may be defined by separation with space, by separation with fire barriers, by diking, by special drainage, or by a combination of these features.

Drainage is desirable to prevent flow between systems. The four methods used for this purpose consists of grading, diking, trenching, and underground or enclosed drains. The degree of hazard will determine the type of drainage or combination of drainage to be used.

Water-spray systems should be hydraulically calculated to deliver the design density at the water supply that is available to the facility. The hydraulic design of the water-spray system is very similar to the design of conventional closed-head sprinkler systems and is done on the basis of determination of friction losses to size the piping. Consideration of the proper discharge pressure, particularly in outdoor installations, must be a factor in the design. This will dictate higher nozzle pressures in outdoor installations to overcome the effect of wind forces.

The location of valves is an important consideration in the design of water-spray systems. Shutoff and water control valves should be located so as to be accessible during a fire, with the water control valve located as close to the hazard as this accessibility will permit. Close accessibility will lessen the quantity of pipe needed and also will ensure as prompt a discharge of water on the hazard as possible.

Explosion Suppression Systems

Many processes involve the potential for an explosion and even though extraordinary precautions are taken, explosions still may occur. In these situations, it may be advisable to contemplate installing an explosion suppression system, depending upon the values involved, the problem with continuity of operations, and other factors.

Three major considerations are necessary in the design of an explosion suppression system:

Detection: Detection must be automatic, extremely quick, and able to distinguish between an explosion and ambient variables. This is normally done with a very sensitive pressure transducer for dust explosions and pressure sensors or ultraviolet radiation detectors for gas explosions. As seen in Figure 10.21, a typical detector operates in 35 msec at 0.20 psi.

Suppression: The extinguishing agent must operate at very high speeds, within a few milliseconds after detection. This is normally accomplished with an electroexplosive release of the agent, which is in a superpressurized state, through the use of high-rate discharge containers (refer to the section on "Halon Systems").

Suppressing Agent: Halons are commonly used as the suppressing agent, with the pressurizing by dry nitrogen. Dry chemical (ammonia phosphate) is also used as an agent. Suppression is basically by chemical inhibition, along with cooling, inerting, or blanketing.

General guidelines for the design, installation, and maintenance of explosion suppression systems can be found in the *Standard on Explosion Prevention Systems*, NFPA 69 (2), the Factory Mutual *Loss Prevention Data Books* (4), the *Fire Protection Handbook* (6), and the *SFPE Handbook of Fire Protection Engineering* (11).

MANUAL SUPPRESSION EQUIPMENT

There are situations where manual suppression will be important, either in conjunction with the operation of automatic extinguishing systems such as sprinklers or where there is an absence of this protection. The type of manual suppression for in-plant use will take the form of standpipe and hose systems and the use of portable extinguishers.

The need for manual suppression is basically twofold. The first and perhaps the primary reason is to extinguish fires in their developing stage, generally with extinguishers. The second reason is to provide a virtually continual source of an extinguishing agent; in this case, water from hose streams.

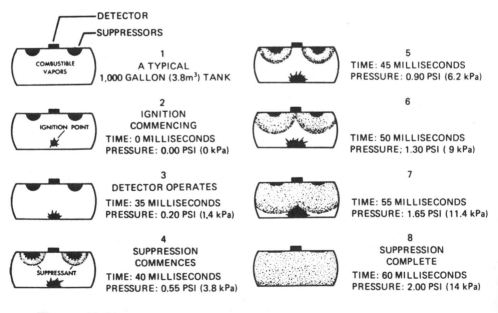

Figure 10.21 Schematic diagram of suppression of an explosion in a typical 1000-gal tank. (Courtesy of Fenwal, Inc.)

Standpipe and Hose Systems

These systems are most frequently used and installed where buildings contain no automatic sprinkler protection, such as large high-rise office buildings. These systems are covered in the *Standard for the Installation of Stand-pipe and Hose Systems*, NFPA 14 (2). In addition to being the primary extinguishing means in buildings, particularly high-rise, small hose systems are essential for the control of flammable liquid spills, for exposure protection on LPG or flammable liquid tanks where the water from a hose will exert a cooling action and hopefully prevent overpressurization, and for the final extinguishment of fires in high-piled rack storage.

Frequently, small-diameter hose is supplied from sprinkler systems from 1-in. or 1.25-in. connections to larger overhead sprinkler members; however, NFPA 14 (2) does not cover this area, which is actually a part of the automatic sprinkler standard, NFPA 13 (2). Before proceeding further, it should be indicated that small-diameter hose refers to 1.5-in. diameter hose, keeping in mind that the hose installed at hydrants or used by fire departments is 2.5-in. diameter.

NFPA 14 (2) defines three classes of service of standpipe and hose systems as follows:

Class I: Systems for use by fire departments and those trained in handling heavy fire streams (2.5-in. hose)

Class II: Systems primarily for use by the building occupants until the arrival of the fire department and involving small-diameter (1.5-in.) hose

Class III: Systems for use either by fire department and those trained in handling heavy hose streams (2.5-in. hose) or by the building occupants (1.5-in. hose)

Although Class I systems are extremely important, most of our attention will be focused on Class II or Class III systems. In any building where a Class I standpipe system is contemplated or required, it would seem reasonable that it would be highly desirable to provide a Class III system, which would permit the use of this system for both heavy fire streams and small-diameter hose.

Standpipes can be either hydraulically designed to provide the required flow at 65 psi or sized as follows. Standpipes for Class I and III service should have minimum 4-in. risers for standpipes not exceeding 100 ft in height and at least 6-in. risers for standpipes in excess of 100 ft in height. Standpipes for Class II service should have a minimum of 2-in. diameter for standpipes not exceeding 50 ft in height and a minimum of 2.5-in. diameter for standpipes exceeding 50 ft in height.

With Class I and III types, minimum flows of 500 gal/min are required where there is only one riser at 65 psi at the topmost outlet. Where there is more than one standpipe or riser involved, the minimum supply of 500 gal/min is expanded to

250 gal/min for each additional standpipe or riser. The water supply for the system should be capable for furnishing a flow for a minimum of 30 min.

Class II service design criteria involving small-diameter hose is considerably less, being 100 gal/min at a residual pressure of 65 psi at the top outlet of each standpipe.

Standpipes are usually wet type, having water pressure maintained at all times and with water available to the valves at hose racks. However, other systems, such as dry type, have no permanent water supply and are generally installed as a time-saving device permitting the fire department to merely hoop up to the fire department connection with hose and charge the line to deliver water to upper stories or inaccessible locations. Standpipes of this type may also be provided with manual operation of valves or remote control devices, or standpipes may be installed so as to automatically admit water into the riser system by opening a valve.

The water supplies for standpipes, may be any of the types of supplies that have been discussed previously for automatic sprinkler systems.

The number of standpipes required is determined to a great extent by the building configuration itself. Standpipe risers should be located, for Class I, II, and III services, so that all building areas are within 30 ft of a nozzle attached to not more that 100 ft of hose. With Class II service, for hose less than 1.5 in. in diameter, each section of a building should be within 20 ft of a nozzle attached to not more than 100 ft of hose. Small-diameter hose can be readily handled by relatively untrained personnel and is installed for both Class II and III services. In the case of Class III services, the small-diameter hose may be from a hose connection separate from the 2.5-in outlet or may be from the 2.5-in. outlet using a reducing coupling.

The hose should be lined, collapsible or uncollapsible. The nozzles and hose rack or reel to which the hose is attached should be of an approved or listed type. It should be recognized that hose on rack must be fully removed from the rack to permit discharge from the nozzle, whereas reels will permit some water discharge without fully removing the hose. There must be a hose valve at the hose rack that is operated manually.

In a typical piping arrangement for a single-zone standpipe system, the connection to the water supply is provided with a shutoff control valve that should be located as close to the supply as possible. In addition, there is a check valve provided in the supply line and in the fire department connection. The standpipes, due to pressures developed, are limited to 275 ft in length. Buildings in excess of this height must be provided with separate zone systems. The hardware necessary for a typical multizone system includes the control valve for each water supply. The valve should be as close to the supply as possible and, in the case of municipal connection, should preferably be a post indicator valve located in the yard to be accessible under emergency conditions. Interior valves should also be located so as to be accessible for operation under adverse conditions. Drains should be provided for each riser, with a drainage system to carry away the water and gauges at each water supply and at the top of the risers.

Portable Extinguishers

The necessity of attacking fires as early as possible can be readily recognized, and it is also advisable to provide an easily handled means of accomplishing this objective. The availability of portable extinguishers is important to provide a means of extinguishing fires before they reach the stage of actuating sprinkler heads or develop, in the case of unprotected facilities, into fires that are beyond the capabilities of the use of portable extinguishers.

As stated in Chapter 3, Elements of a Fire Loss Control Program, OSHA regulations (29 CFR 1910.157) on portable fire extinguishers apply when the extinguishers are provided for use by the employees (3).

The effectiveness of fire extinguishers, as in the *Standard for Portable Fire Extinguishers*, NFPA 10 (2), depends upon the following:

1. Proper location of extinguishers
2. Proper working order of extinguishers
3. Proper types of extinguishers for hazards encountered
4. Discovery of the fire while it is still small enough for the fire extinguisher to be effective
5. Availability of persons able to use and handle extinguishers

In determining the type of extinguishers to be provided for a particular hazard, we must remember the classes of fires, A,B,C, and D. Class A fires are those in ordinary combustible materials such as wood and paper. General distribution is on the basis of extinguishers needed for Class A-type hazards, with these types of extinguishers consisting basically of water type, although AFFF and multipurpose dry-chemical extinguishers can be used.

Extinguishers are classified by letter and numeral, with the letter indicating A, B, C, or D for the class of fire for which the extinguisher will provide an effective agent and, in the case of Class A and B extinguishers, a numeral indicating the relative effectiveness of the extinguisher. For example, a 2A-rated extinguisher has twice the effectiveness of a 1A-rated extinguisher. No numerical designation is made for Class C-type extinguishers, which may be used on energized electrical equipment where the extinguishing agent obviously is nonconductive, or for Class D extinguishers, which are for use on combustible metal fires. Class D extinguishers will have the relative effectiveness of the extinguishers for use on specific combustible metal fires detailed on the extinguisher name plate.

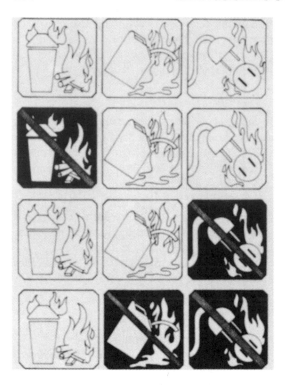

Figure 10.22 Recommended pictographs. When an application is prohibited, the background is black and the slash is bright red; otherwise the background is light blue. Top row indicated an extinguisher for Class A,B, and C fires; second rows, Class B and C fires; third row, A and B fires; and bottom row, Class A fires. (Reprinted with permission from the *Fire Protection Handbook*, 16th Edition, Copyright 1986, National Fire Protection Association, Quincy, MA., 02269.)

For marking extinguishers, the most recent recommended system is pictographs of both use and nonuse on the label, as shown in Figure 10.22. Letter shaped symbol markings (green triangle for Class A; red square for Class B; blue circle for Class C; and yellow star for Class D) are still being used in the interim.

As in many areas of fire protection, to properly determine the numbers required, after it has been determined what extinguisher is necessary to combat the hazard involved, it then must be decided whether the hazard is light, ordinary, or extra hazard. these hazards are defined in NFPA 10 (2) as follows:

Light: With the amount of combustibles or flammable liquids present, fires of small size may be anticipated. This class will include offices, school areas, assembly areas, churches, etc.

Ordinary: With the amount of combustibles or flammable liquids present, fires of moderate size may be anticipated. These may include mercantile storage an display, parking garages, light manufacturing, warehouses not classified as extra hazards, etc.

Extra: With the amount of combustibles or flammable liquids present, fires of severe magnitude may be anticipated. These may include woodworking, warehouses with high-piled storage, areas involving processing or storage or flammable liquids, etc.

NFPA 10 (2) should be consulted for tables indicating the extinguisher area requirements and maximum travel distance for the various occupancy hazard classification. The maximum travel distance from any point to an extinguisher is 75 ft for Class A hazards and 30 or 50 ft, depending on the size of the extinguisher, for Class B hazards.

SUMMARY

This chapter has been designed to acquaint the manager with the types of systems available for the protection of the employees and facilities; their advantages and limitations; and, where appropriate, some design criteria. Additional sources of information generally have been given for those who wish to learn more.

REFERENCES

1. *The Foundations of Loss Control*, Factory Mutual Engineering Corporation, Norwood, Mass., 1987.
2. *National Fire Codes*, National Fire Protection Association, Quincy, Mass., 1989.
3. *General Industry Standards—29 CFR 1910*, Occupational Safety and Health Administration, Washington, D.C., 1988.
4. *Loss Prevention Data Books*, Factory Mutual Engineering Corporation, Norwood, Mass., 1989.
5. Casaccio, Ellen K., "Research into the Nineties Sprinkler Technology," *Record*, 66 (3): 9–12 (1989).
6. Cote, Arthur E., ed., *Fire Protection Handbook*, 16th Edition, National Fire Protection Association, Quincy, Mass., 1989.
7. *Fire Protection Equipment Directory*, Underwriters laboratories, Inc., Northbrook, Ill., 1989.
8. *ASME Boiler and Pressure Vessel Code*, American Society of Mechanical Engineers, New York, 1977.
9. "Facts about Halon," The Halon Research Institute, Chicago, Ill., 1988.

10. "Halon Fire Suppressants Update," *Conservation Conversation*, 10 (2): 1, 2 (1989).
11. *SFPE Handbook of Fire Protection Engineering*, 1st Edition, National Fire Protection Association, Quincy, Mass., 1988.

11

Automatic Fire Detection and Alarm Systems

An automatic fire detection and alarm system (AFDAS) is one of the fundamental elements in the overall fire protection of any facility or plant. An AFDAS can help limit fire losses, indicating loss of life, in buildings of all occupancies. Looking at fire incident data from the National Fire Incident Reporting System of the National Fire Data Center, over 90% of the fire incidents with casualties and approximately 88% of the worker injuries occurred with no detectors present (1, 2). To provide the greatest benefit, an AFDAS should perform several different functions: (1) be reliable and detect fires while rejecting extraneous signals; (2) initiate an alerting signal to building occupants; and (3) initiate an identifying signal to the responsible people (fire department, control center, etc.) and/or to automatic suppression systems (3, 4).

AFDAS may be installed for a number of reasons. Depending on the type of system, they can be used for early warning of a fire condition and can sound an alarm for evacuation. Signals can be transmitted to a public fire department or fire brigade to summon help.

AFDAS are required under certain OSHA regulations—29 CFR 1910.164 (5) for fire detection systems and 1910.165 for employee alarm systems. OSHA requires that employees "establish procedures for sounding emergency alarms in the workplace." This can range from simple alerting by direct voice communication (employers with 10 or fewer employees) to a complex AFDAS.

One of the more frequent uses of AFDAS will be automatic detection as a most important component of automatic protection systems, including sprinkler, foam, water spray, carbon dioxide, halon, dry chemical, and explosion suppression systems. These systems were covered under Chapter 10, emphasizing automatic detection to provide the means of alarm and actuation.

AFDAS may be extremely simple, as typified by a fusible link release system for dry-chemical protection in a cooking hood or for the release of an automatic smoke and heat vent. They also may be highly complicated systems involving sophisticated detection equipment such as smoke and flame detectors and including, supervised circuits.

SYSTEM DESCRIPTION

All AFDAS have the following basic features:

1. Control/fire alarm panels with primary and secondary power supplies (process function in Fig. 11.1)
2. Sensor/initiating devices (detect function in Fig. 11.1)
3. Alarm/indicating devices (alert and alarm functions in Fig. 11.1)

The simplest system is the single station detector, which consists of the detection device, a simple electronic or mechanical processor, and a bell or horn, all in common housing. This system provides detect and alert functions, but not alarm or optional control functions (see Fig. 11.1). For industrial and commercial occupancies, AFDAS become more organized and elaborate (3).

SYSTEM COMPONENTS

AFDAS are traditionally classified by the method of transmission of a remote alarm and the nature of the individuals or equipment to which the alarm is transmitted. These classifications—local, auxiliary, remote station, proprietary, and central station signaling systems—will be discussed later in this chapter. Some components are common to all alarm systems and are compiled in Table 11.1.

Automatic Fire Detection Devices

Automatic fire detection devices are one of the key detecting and initiating components of a AFDAS. There is a wide variety of detection devices that are actuated by the elements of a fire, the most common being heat, smoke, and light radiation (flame) (3, 6).

When determining what detector should be used, the most important consideration, other than environmental factors, is the type of occupancy. This includes the energy levels anticipated, the type of fire and the quantities of smoke that may be developed, and the rate of heat release. For example, a properly installed and positioned heat detector will respond rapidly to a high-energy-release fire, such

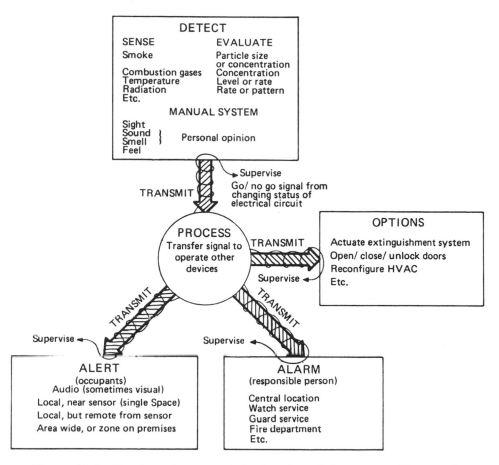

Figure 11.1 Functions of an automatic fire detection and alarm system. (Courtesy of National Institute for Occupational Safety and Health.)

as created by ignition of flammable liquid vapors. The same detector, however, would not be very sensitive to a low-energy fire, such as might be expected in electronic equipment typically found in control rooms and data-processing areas, or to a low-energy smoldering fire. While heat detectors will respond to high-energy-release fires, more sophisticated flame response detection equipment is now available that will alarm more rapidly to this type of fire.

Several factors influence the location and spacing of detectors. As a starting point, one should consider the linear maximum spacing as specified in listings of

Table 11.1   Alarm System Components[a]

COMPONENT	DESCRIPTION
1. *CONTROL AND STATUS EQUIPMENT*	
Control Panel	Provides enclosures that house the primary process function, an assembly of electrical components. Has provisions for the connection of signal initiation devices, signal indicating devices, and power to form a coordinated system. Usually provides capability to supervise system integrity. Main panels with several remotely located subpanels are a common system design where necessary.
Power Supply	Provides independent sources of main power, trouble power, and emergency power of sufficient capacity to cause all operations of the protective alarm system to function.

Annunciator	Produces a visual display of the signals present in the system circuitry:
	 1. Illuminated
	     a) Bull's eyes (a group of individual indicating lamps with colored jewels).
	     b) Windows (a group of black lighted windows whose message is illuminated when energized).
	     c) Graphic (displays a plan of the area covered; may have bull's eyes or windows).
	 2. Target or Drop (an electromechanically operated device that displays an indicator when operated).

Multiplexing Equipment	Electronic coding of multiple signals on a common circuit. Devices such as controllers, transmitter/receivers, central minicomputers, and CRT terminals are necessary for control and status indication in a multiplex system.
Line Printer	Records all events, acknowledgments, system status changes, operator actions, etc., that occur in a large fire alarm system.

2. *INPUT HARDWARE*

A. Alarm Initiating Devices

Automatic Fire Detection Devices	Devices designed to automatically respond (sense, evaluate) to one or more fire signatures with a signal or relay closure; e.g., smoke detectors, heat detectors, etc.
Manual Fire Detection Devices	Devices designed to respond to manual stimuli with a signal; e.g., manual pull stations, call stations, etc.
Sprinkler System Waterflow Alarm Devices	Devices designed to automatically introduce a signal into the alarm circuitry when its vane has been deflected by a flow of water in the sprinkler system.
Intrusion and Burglary Detection Devices	Devices designed to automatically introduce an intrusion signal into the system circuitry. These devices include door and window monitors, photoelectric and ultrasonic intrusion mechanisms, etc.

B. Supervisory Devices[b]

Sprinkler System Supervisory Devices	Devices designed to automatically introduce a supervisory signal into the system circuitry when some component of the sprinkler system is not in the operating mode, such as: gate valve supervisory switches, post indicator valve supervisory switches, water tank level switches, etc.
Hazardous Equipment Supervisory Devices	Devices designed to automatically introduce a supervisory signal into the system circuitry when some piece of equipment reaches a dangerous operating level of temperature or other condition, such as tank temperature switches.

Table 11.1 (Continued)

COMPONENT	DESCRIPTION
3. *OUTPUT HARDWARE*	
A. Alarm Indicators	
Audible Devices	Any noise-making device used to indicate trouble or an emergency, such as bells, gongs, chimes, buzzers, horns, sirens, howlers, and trumpets.
Visual Devices	Any attention-getting visual device used to indicate trouble or an emergency, such as flashing lights, normally used where deaf people are present or the ambient noise levels are high.
Alarm Relay Devices	Devices designed to automatically relay the general building fire alarm to another location by means of an auxiliary connection, remote station connection, central station connection, or proprietary connection.
B. Equipment Control	
Automatic Smoke Control Devices	Devices designed to control smoke in a building by diluting air, exhausting air, and confining air, by use of dampers, door releases, and other HVAC devices. These devices can be integrated into the fire alarm system in a custom fashion. For example, detector activation of a total flooding Halon 1301 system should simultaneously activate dampers, closing all vents to the room confining air.
Suppression System Activation Devices	Devices designed to automatically activate fire suppression systems, upon signaling from the control panel, which is relayed from some specific fire detectors. These devices include solenoid valves, electroexplosive devices, etc.
Security Access Devices	Door openers, etc., can be integrated into the alarm system when desired.

[a]Courtesy of National Institute for Occupational Safety and Health.
[b]The reader should note that, unfortunately, the term "supervisory" is in common use to describe two different system functions. The first is system supervision of its own circuit integrity. The second type of supervision is the option of monitoring plant processes, readiness of suppression systems, etc.

detectors in UL's *Fire Protection Equipment Directory* (7) and Factory Mutual's *Approval Guide* (8) for the particular device desired. Other factors to consider are

1. Building construction and ceiling obstructions
2. The susceptibility and burning characteristics of the contents involved
3. Airflows throughout the area involved

The building construction and particularly the type of ceiling affect the spacing of detectors. Tests and maximum spacing requirements of the approval agencies are based upon smooth ceilings. Ceilings with projections such as beams and girders, may require considerably closer spacing than the maximum permitted by the actual listings.

Perhaps the major consideration for location of detectors, particularly in facilities, such as computer areas, that are subject to above-average air changes and air velocity, is the airflow. The improper location of a smoke detector near the air supply can prevent alarming; the proper location is near the air return. A detector located near the air supply will not alarm, because the flow of any combustion particles will be away from the detector and toward the return, with the air supply considerably diluting the air contaminant.

Closer spacing is a must for proper area protection in locations having high airflows, since there is considerable dilution. Airflows can be of considerable significance in the location of all types of detectors, including heat detectors and smoke detectors, because both heat and smoke will have a tendency to be conveyed by airflows.

With regard to air-conditioning systems, in-duct smoke detectors are often required by code to provide for alarm and/or to shut down the fans to prevent the circulation of smoke and heat throughout the building. This type of installation should not be considered an acceptable alternative to area detection, because of unreliability involving the dilution of smoke and combustion contaminants by the large quantities of air moved through the systems. For in-duct detection, either photoelectric or ionization detectors are used (9,10,11).

From the viewpoints of life safety and protection of valuable equipment, a greater degree of protection is needed, requiring a type and spacing of detectors that will be the most sensitive to the type of fire that might be anticipated.

Each detector installation should be designed on the basis of a complete engineering survey conducted by competent personnel in this field. The desirability of such a survey cannot be overemphasized, and any manager contemplating the installation of detectors should insist upon it. This survey must begin with an analysis of the hazards and susceptibility of equipment. Further reference should be made to the Standard on *Automatic Fire Detectors*, NFPA 72E(9), for the proper location, testing, and maintenance of fire detectors. The most common types of detectors are discussed in the following sections.

Heat Detectors

Heat detectors respond when the detecting element reaches a predetermined fixed temperature or a specified rate of temperature change. A common sprinkler is an example of a combined heat detector and extinguishing device when the system has water-flow indicators. Heat detectors are one of the most common types of detectors used and are the least expensive. These detectors also have the lowest false alarm rate compared to other types of detectors (3, 6, 12).

Fixed-Temperature Types   Fixed-temperature heat detectors are designed to respond to a specified temperature, ranging from 135 to 575°F (57 to 302°C). They are good for general spot and small area detection and have a high degree of reliability.

The fusible-element type uses a eutectic metal (alloys of bismuth, lead, tin, and cadium) that melts rapidly at a predetermined temperature. The alarm is activated by the fusing of the metal, which either opens a circuit or releases a spring under tension (Fig. 11.2). This principle is also commonly found in sprinkler head elements and in fusible links for releasing devices.

The bimetallic type uses two metals bonded together, each with a different coefficient of thermal expansion. When heated, the differences in expansion cause the metal to flex and close a contact point. This type of detector is self-restoring. Two common designs are the bimetal strip (Fig. 11.3) and the snap disc (Fig. 11.4). The rate-compensation detector (Fig. 11.5) uses a metal cylinder containing two metal struts that are the alarm contacts. During a rapid increase in temperature, the cylinder expands faster than the struts, thereby closing the struts. Under slow temperature buildup, both the cylinder and the struts expand until the contacts close. In this case, the cylinder and the struts are the two different metals. This type of detector compensates for the thermal lag (3, 6, 12).

Certain types of rate-compensation detectors can be used in Class I and II hazardous areas. Occasionally, this type of detector is used as a second detection system to actuate a halon system when the prealarm is by a smoke detector.

The continuous-line type has two basic designs. The first uses a pair of wires in a normally open circuit. Conductors are insulated from each other by a thermoplastic of known fusing temperature, twisted and installed under tension (Fig. 11.6). Actuation occurs when the insulation melts, closing the contacts. This type is not self-restoring and is used where continuous-line heat detection is desirable.

Another design uses a stainless steel tube with a coaxial center conductor (Fig. 11.7). Normally, a small current flows through the conductor, but under fire conditions, the resistance decreases, allowing a greater current to actuate the alarm. This type is self-restoring and has had an interesting application in cable trays carrying numerous power and/or control cables to important process or communications equipment where fire in the tray could create possible long-term business interruption problems (3, 6).

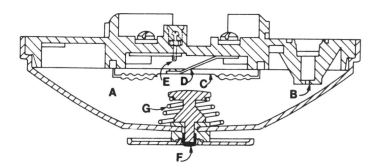

Figure 11.2 Cross section of a fixed-temperature/rate-of-rise heat detector. Air in chamber A expands more rapidly than can escape from vent B, causing pressure to close contact D between diaphragm C and contact screw E. Fixed-temperature operation occurs when fusible alloy F melts and releases spring G, which presses the diaphragm C, closing the contacts. (Courtesy of the Edwards Company.)

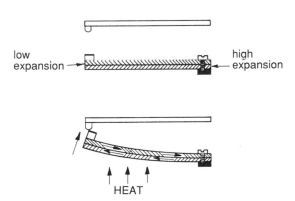

Figure 11.3 Bimetal-strip heat detector principle. (Courtesy of U.S. Department of Commerce.)

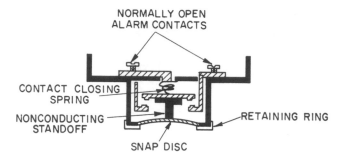

Figure 11.4 Schematic of snap-disc heat detector. (Courtesy of U.S. Department of Commerce.)

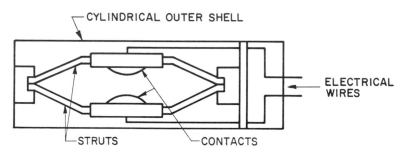

Figure 11.5 Schematic of a rate-compensation heat detector. (Courtesy of U.S. Department of Commerce.)

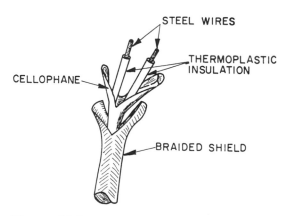

Figure 11.6 Line-type detection cable. (Courtesy of U.S. Department of Commerce.)

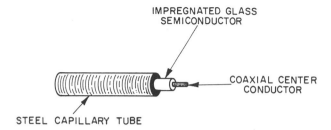

Figure 11.7 Line-type fire detection cable using a glass semiconductor. (Courtesy of U.S. Department of Commerce.)

Some newer developments in the fixed-temperature type are the capacitor cable, the synthetic filament transducer, and metal oxide thermistors.

Rate of Temperature Rise Types   The rate of temperature rise or rate-of-rise detectors actuate when the rate of temperature rise exceeds around 12 to 15°F (7 to 8°C) per minute. Ambient temperature changes are compensated for within the detector. These detectors are good for general spot and area detection and are typically faster than fixed-temperature heat detectors (3, 6, 12).

In the pneumatic type, air expands when heated in a closed system, generating a mechanical force that closes contacts to activate the alarm. The spot-type pneumatic detector uses heat-collecting air chambers (Fig. 11.2) and the line type consists of metallic tubing in a loop configurations attached to the ceiling (Fig. 11.8).

Thermoelectric-type detectors use thermocouples to produce an increase voltage with an increase in temperature. When the voltage increases at an abnormal rate, the detector actuates.

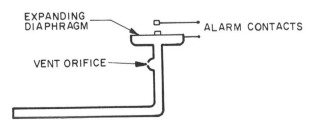

Figure 11.8 Line-type rate-of-rise heat detector principle. (Courtesy of U.S. Department of Commerce.)

Some heat detectors use a combination of types. For instance, the fixed-temperature/rate-of-rise detector shown in Figure 11.2 takes advantage of both principles.

Smoke Detectors

Ionization Type   In ionization-type smoke detector reacts to the aerosol components of combustion, responding best to particle sizes in the range of 0.01 to 1.0 micrometer and with a normal sensitivity range of 0.5% to 3.5% per foot obscuration. The basic detection mechanism consists of alpha or beta radiation source in a chamber containing positive and negative electrodes (Fig. 11.9). The radiation source ionizes the oxygen and nitrogen molecules in the air between the electrodes, which causes a small current flow. When aerosols and smoke enter the chamber, ion mobility is decreased and resulting decreased in current actuates the alarm. This detection is good for general spot and area detection. Most units have adjustable sensitivity within a certain range. The sensitivity may be affected by

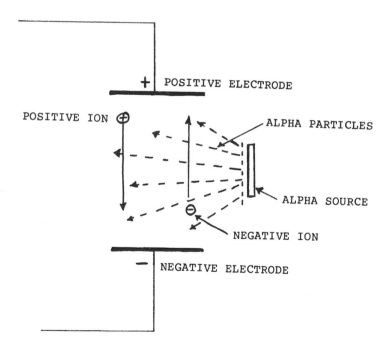

Figure 11.9   Ionization-type smoke detector principle.

Figure 11.10 Beam-type smoke detector principle. (Courtesy of U.S. Department of Commerce.)

air velocity entering the chamber, humidity, altitude, and high ambient radioactive levels. The detector is self-restoring and tends to be more sensitive to flaming fires than to smoldering fires (3, 12, 13, 14).

Photoelectric Type The photoelectric-type smoke detector reacts to aerosol components of combustion, responding best to particles greater than 0.5 micrometer. The basic detection mechanism consists of smoke affecting the propagation of light as it passes through the air, with two methods, line type or light-scattering type. The line type consists of the attenuation of the light intensity integrated over the entire beam path length with a light source, a collimating lens system, and a photosensitive cell (Fig. 11.10). With this type, the smoke obscures the light beam, thereby activating the alarm. This type is self-restoring and is good for large open-area detection.

The light-scattering type scatters light in the forward direction and at various angles to the light beam path. Smoke particles enter a normally dark chamber and scatter light onto the photosensitive cell, activating the alarm (Figs. 11.11 and 11.12). The normal range of sensitivity is 0.5 % to 2.5% light obscuration per foot. This type provides general spot and area detection is self-restoring and tends to be more sensitive to smoldering fires than to flaming fires (3,12,13,14).

Condensation Nuclei Type The condensation nuclei (cloud chamber) type of smoke detector uses a technique in which particles (0.001 to 0.1 micrometer) are made to act as condensation nuclei to form a cloud in a chamber within the detector. The density of the cloud is then measured by a photoelectric principle, and the detector responds at a predetermined level. This detector samples the air from the protected areas by intermittent sampling in a multizone system and is used in high-valued areas, such as museums, art galleries, etc. This is an extremely sensitive detector with a wide range of sensitivity (3, 12).

Fire Gas Sensing Detectors

Fire gas sensing detectors normally use the semiconductor principle or the catalytic element principle to detect combustible gases during a fire. The detector is normally used for spot and area detection and is self-restoring.

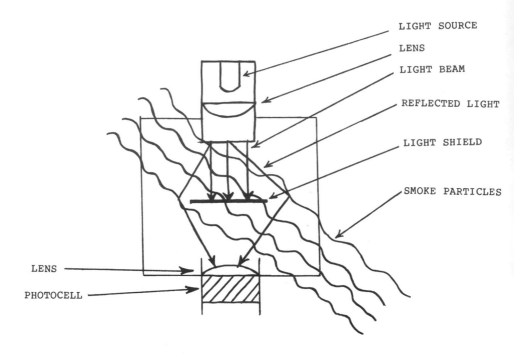

LIGHT SOURCE

LENS

LIGHT BEAM

REFLECTED LIGHT

LIGHT SHIELD

SMOKE PARTICLES

LENS

PHOTOCELL

Figure 11.11   Photoelectric light-scattering (light shield) type principle.

Flame Detectors

Flame detectors operate by sensing the radiant energy from a fire. The two basic
types are infrared (Fig. 11.13) and ultraviolet (Fig. 11.14). The detectors are used
in general area detection, are self-restoring, and have a high degree of sensitivity
and speed of response. Flame detectors are generally used in high-hazard areas,
such as fuel loading platforms, hyperbaric chambers, industrial process areas, and
areas where explosive atmospheres may occur (3, 12).

SYSTEM CLASSIFICATIONS

As stated earlier, AFDAS are traditionally classified by the method of transmission of a remote alarm and the nature of the individuals or equipment to which the
alarm is transmitted. There are basically five signaling systems: local, auxiliary,
remote station, proprietary, and central station.

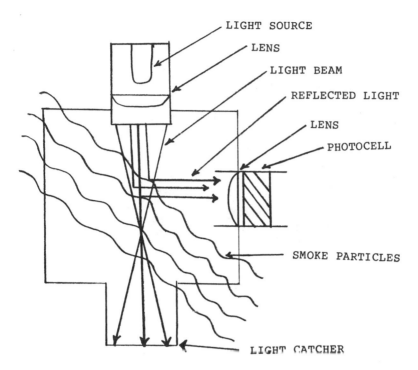

Figure 11.12   Photoelectric light-scattering (light at right angle) type principle.

Local System

The local protective signaling system has as its main purpose the local evacuation or alarm of personnel solely within the protected building (Fig. 11.15). An example of a local system is an area detection system located in an office that, upon alarming, produces an audible signal only at the site rather than transmitting that signal to a constantly attended location either on or off premises. The full requirements of local systems are found in the *Standard for the Installation, Maintenance and Use of Local Protective Signaling Systems for Guard's Tour, Fire Alarm and Supervisory Service*, NFPA 72A (9). Such a system providing for sprinkler water flow, along with sprinkler control valve tamper switches, would be anticipated to sound an alarm on the premises or to operate an annunciator panel located on the premises, but not necessarily constantly attended.

All systems of this type should be electrically supervised. This means that in the event of a fault condition consisting of either an open circuit or a ground that will prevent the required operation of the signal, a distinctive trouble signal will

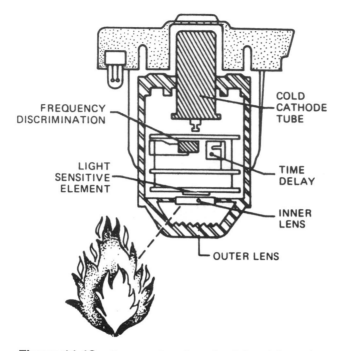

FREQUENCY
DISCRIMINATION

COLD
CATHODE
TUBE

LIGHT
SENSITIVE
ELEMENT

TIME
DELAY

INNER
LENS

OUTER LENS

Figure 11.13 Cross-section view of an infrared flame detector. (Courtesy of Cerberus Pyrotronics.)

be furnished. This should include supervision of the power supply circuits, circuits for signals initiated by the operation of the fire detectors or other equipment, and alarm signal sounding circuits. The distinctive trouble signal should be indicated by continuous operation of a sounding device.

Only one main power supply is required for a local system; however, a secondary source of power is encouraged and, when provided, should operate the system for a period of 24 hr and then operate the alarm system for 5 min (covers the normal day of occupancy with a 5-min alarm for evacuation) (6,9).

Auxiliary System

The auxiliary protective signaling system provides protection by transmitting an alarm to the fire department over the municipal fire alarm system, either through a nearby master fire alarm box or through a dedicated telephone line directly to the fire department communication switchboard (Fig. 11.16). Auxiliary systems are required to have a primary power supply for operation and: (1) a secondary or standby power source for operation of the system for 60 hr followed by 5 min of

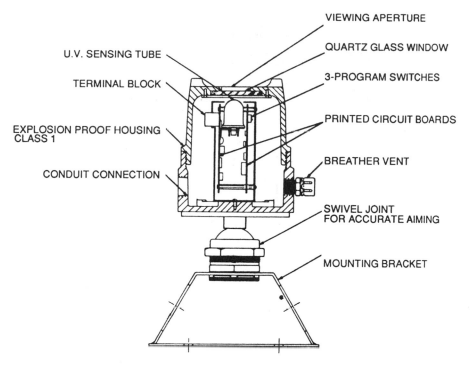

Figure 11.14 Cross-section view of an ultraviolet flame detector. (Courtesy of Alison Control, Inc.)

alarm, and (2) a trouble signal supply for operation of trouble signals, except that standby power may be used for this purpose (9).

Electrical supervision much the same as that for local systems is required. Auxiliary alarm systems are described in depth in the *Standard for the Installation, Maintenance and Use of Auxiliary Protective Signaling Systems for Alarm Service*, NFPA 72B (9).

Remote Station System

Remote station protective signaling systems are systems where signals are received at locations where qualified personnel are in constant attendance to receive and recognize such signals (Fig. 11.17). Normally, remote alarm systems include termination of signals at fire stations or other governmental agencies or a telephone answering service using leased telephone lines. The power supply sources are required to be the same as for auxiliary signaling systems as described above and are more fully described in the *Standard for the Installation, Mainte-*

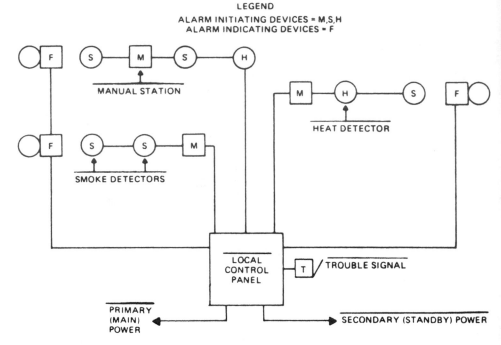

Figure 11.15   Typical local fire alarm system. (Reprinted with permission from the *Fire Protection Handbook*, 16th Edition, Copyright 1986, National Fire Protection Association, Quincy, MA 02269.)

nance and Use of Remote Station Protective Signaling System, NFPA 72C (9). Electrical supervision of the system is also similar to that of auxiliary alarm systems.

Proprietary System

The proprietary protective signaling system is a widely used type of system in large industrial and commercial facilities. A proprietary signaling system transmits a signal to a central supervising station at the protected property, which should be attended by competent personnel at all times (Fig. 11.18). For full requirements of proprietary systems, refer to the *Standard for the Installation, Maintenance and Use of Proprietary Protective Signaling Systems*, NFPA 72D (9).

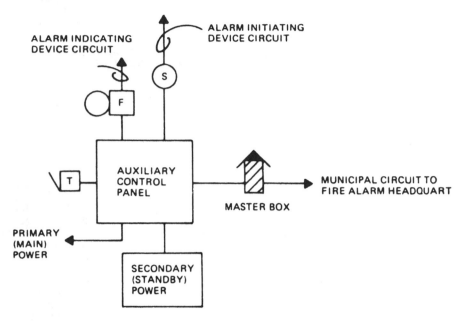

ALARM INITIATING
DEVICE CIRCUIT

ALARM INDICATING
DEVICE CIRCUIT

S

F

T

AUXILIARY
CONTROL
PANEL

MASTER BOX

MUNICIPAL CIRCUIT TO
FIRE ALARM HEADQUART

PRIMARY
(MAIN)
POWER

SECONDARY
(STANDBY)
POWER

Figure 11.16 Typical auxiliary fire alarm system. (Reprinted with permission from the *Fire Protection Handbook*, 16th Edition, Copyright 1986, National Fire Protection Association, Quincy, MA 02269.)

A complete proprietary system will have a main power supply source and a secondary power source (period of 24 hrs) within the supervisory station and will also be electrically supervised.

Newer systems have signal multiplexing and built-in minicomputer systems. The system receiving console consists of individual lights, a digital display, or a CRT (cathode-ray tube) visual display indicating the alarm point, with a audible alarm to alert the console operator and a hard-copy printer. Large systems usually do more than indicate fire alarms. The systems can provide for smoke control, HVAC (heating, ventilating, and air conditioning) system controls, other system controls, and energy management capabilities. This versatility has been a major factor in the increased use of this type of system in larger facilities (6,11).

Central Station System

The most commonly used signaling system, other than local system, is the central station protective signaling system. It provides for the transmission, receiving, and recording of signals at an approved central station in which competent and experienced observers and operators are in constant attendance to take required

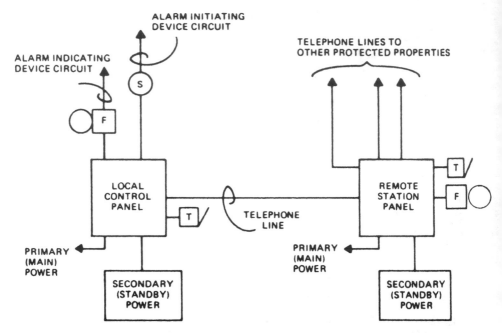

Figure 11.17   Typical remote station fire alarm system. (Reprinted with permission from the *Fire Protection Handbook*, 16th Edition, Copyright 1986, National Fire Protection Association, Quincy, MA 02269.)

action upon receipt of a signal (Fig. 11.18). This system is similar to a proprietary system. However, there are differences in the way that signals are transmitted between the facility and the receiving station.

The central station system uses the McCulloh circuit, which normally transmits over two wires, but can be switched manually or automatically to transmit over one wire and ground. In this way, the system will not be made inoperative by a break or ground fault in a single line (6).

Additional requirements can be found in the *Standard for the Installation, Maintenance and Use of Central Station Signaling Systems*, NFPA 71 (9). As in some of the other systems, both primary and secondary power supplies are required, with a separate supply for the trouble signal, except that the secondary supply may be used for this purpose. This system is also required to have electrical supervision to provide for a distinctive trouble signal in the event of a break or ground fault condition.

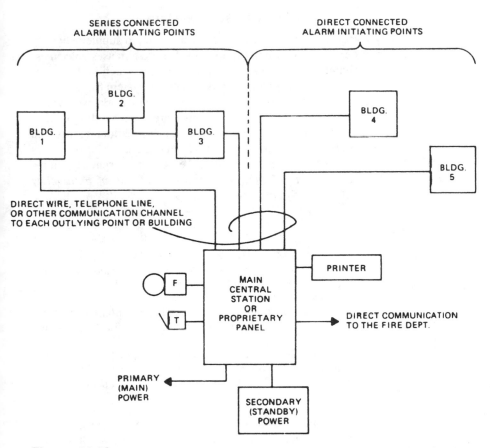

Figure 11.18 Typical proprietary or central station fire alarm system. (Reprinted with permission from the *Fire Protection Handbook*, 16th Edition, Copyright 1986, National Fire Protection Association, Quincy, MA 02269.)

When a signal is received at the central station, the fire department and other appropriate authorities are alerted. In addition, depending on the type of alarm, a runner is dispatched to the protected facility.

Frequently, false signals will be received from centrally supervised sprinkler water flows where there are rather extreme fluctuations in pressure. To overcome this, excess pressure supervisory pumps are provided on the systems. These are nothing more than small-capacity pumps that provide high pressure on the system to prevent surges from initiating a signal to the central station.

Emergency Voice/Alarm Communication System

The emergency voice/alarm communication system is used to supplement a local, auxiliary, remote station, or proprietary system. This system uses voice messages over a high-reliability speaker system to give special instructions to the occupants. The system requires standby power to operate the system for 24 hr. followed by 2 hr. of voice/alarm signaling. This type of system works well where occupants need to be instructed to go to a safe refuge area of a building if complete evacuation is not feasible. Communication can then be maintained with the occupants to prevent panic and facilitate further relocation. Survivability for this type of system is very critical and reference should be made to the special requirements in the *Standard for Installation, Maintenance and Use of Emergency Voice/Alarm Communication Systems*, NFPA 72F (9).

REFERENCES

1. Bochnak, Peter M. and Moll, Michael B., "Relationships Between Worker Casualties, Other Fire Loss Indicators, and Fire Protection Strategies" (internal report), U.S. Department of Health and Human Services, National Institute for Occupational Safety and Health, Morgantown, W. Va., 1984.
2. Federal Emergency Management Agency, U.S. Fire Administration, National Fire Incident Reporting System Unpublished Fire Incident Data for 1977 through 1980.
3. Mniszewski, K., Wakeley, H., Waterman, T., and Campbell, J., "State of the Art: Fire Alarm Technology Related to Protecting Life in Work Places,"U.S. Department of Health and Human Services, National Institute for Occupational Safety and Health, Morgantown, W. Va., 1978.
4. Bochnak, Peter M., "State of the Art: Fire Alarm Technology Related to Protecting Life in Workplaces," Proceedings from the 52nd Fire Department Instructors Conference, The International Society of Fire Service Instructors, March 24 en 27, 1980, Memphis, Tenn.
5. *General Industry Standards—29 CFR 1910*, Occupational Safety and Health Administration, Washington, D.C., 1988.
6. Cote, Arthur E., ed., *Fire Protection Handbook*, 16th Edition, National Fire Protection Association, Quincy, Mass., 1986.
7. *Fire Protection Equipment Directory*, Underwriters Laboratories, Inc., Northbrook, Ill., 1989.
8. *Approval Guide*, Factory Mutual Engineering Corporation, Norwood, Mass., 1989.
9. *National Fire Codes*, National Fire Protection Association, Quincy, Mass. 1989.
10. *SFPE Handbook of Fire Protection Engineering*, 1st Edition, National Fire Protection Association, Quincy, Mass., 1988.
11. *The Handbook of Property Conservation*, 3rd Edition, Factory Mutual Engineering Corporation, Norwood, Mass., 1983.

12. Custer, Richard L. P. and Bright, Richard G., *Fire Detection: The State-of-the-Art*, U.S. Department of Commerce, National Bureau of Standards, Washington, D.C., 1974.
13. Bochnak, Peter M., "Developments in Fire Detection," Proceedings from the Mutual Engineers' Conference, American Mutual Insurance Alliance, April 5 – 7, 1976, Atlanta, G.
14. Bochnak, Peter M., "Smoke and Fire Detectors for Wood Heating Systems," Proceeding from the Wood Heating Seminar I, Wood Energy Institute, April 20 – 21, 1977, Cambridge, Mass.

12

Education and Training

As stated in earlier chapters, training is necessary for emergency organizations, emergency action procedures, emergency equipment, and general fire prevention. The aim of training is to create an attitude of fire safety in the facility and recognition by all employees that protecting the facility from fire also protects their lives and livelihood. The human factor (refer to Chapter 3) plays a major role in the loss control program, and training plays a vital role to help create the desired fire-conscious attitude in employees. A clear commitment by the company to become more safety conscious must be reflected in the training programs (1, 2).

DEVELOPING A FIRE SAFETY PROGRAM

There are several steps that can be used to develop a fire safety program. These steps are rather general, but give a framework of the logical sequence that can be followed (3).

Identify Training Needs

Training programs should be established for "real" training needs, not just to establish "a program." Unless you have a need for the training, use of person-hours and funds to establish a program can not be easily justified. Training should be aimed at equipping people to perform the duties and functions necessary to accomplish the job. Once the needs are identified, you can begin to establish the objectives of the training.

Training Objectives

The training objectives should be a description of the performance you want trainees to exhibit before considering them competent. The objective describes the intended result of training, rather than the process of training. For example, the statement, "To demonstrate the proper use of fire extinguishers," describes what the trainer will do, not the result of training. The proper training objective would read: "Be able to properly use fire extinguishers on four different types of fires." This gives the result of the training. Objectives in this sense are useful in pointing to the content and procedures that will lead to successful training. Objectives should be developed to meet the training needs (3, 4).

Develop Course Outline

The next step is to develop a training outline. This should start as a list of subjects. For each subject, an outline can be prepared from which one can begin to collect materials (textbooks, reports, videos, etc.).

Training Techniques

From the subject outlines, lesson plans can be developed. These lesson plans should state the specific objective of each subject in terms of desired trainee behavior, rather than instructor performance. The type of teaching method should be decided: lecture, discussion, demonstration, or a combination of these.

Conducting the Program

The logistics of the training program must be considered in order to have an organized program. A training timetable needs to be established. Records, certificates, facilities, evaluations, and equipment must be considered. A number of things should be checked: communications, objectives, training of instructors, and effect on production.

Evaluating the Program

The best way to evaluate the effectiveness of the training program is to check over "a reasonable length of time if you have met the need or solved the problem that existed when you started the program" (3).

EMERGENCY ORGANIZATION AND FIRE BRIGADES

As discussed in Chapter 5, personnel involved with the facility emergency organization should be trained in operating protective equipment, reporting emergencies, responding to fires, fire attack procedures, and operating suppression equipment. Personnel should receive specialized training if they are involved with sprinkler valve control, fire pump operation, electrical safety, and steam and gas piping control. As stated in Chapter 3, the detail of the training will depend upon whether employees will fight fires or not. If all employees or selected employees will fight incipient fires, then they must be trained to use fire extinguishers. If a fire brigade will be trained to used to fight incipient or interior structural fires, they will need additional training as a fire brigade (5, 6).

The training program must be tailored to each plant or facility's needs. In general, training will include instruction in the following areas (2).

Sounding the alarm: All employees should be trained in the correct procedure for reporting an emergency. Designated emergency team members should take part in practice in practice alarm drills.

Response and fire attack: Once the alarm sounds, the designated group should know where to respond. This may be a particular fire area or an assembly point. Periodic drills should be conducted.

Hose handling: Hands-on training should be conducted annually in the use of 1 – 1/2 – in. hose line and small hose lines (5/8 in. to 1 – 1/2 in.). This will develop familiarity with the location and operation of hose racks and reels and with proper care and storage of hose. At larger facilities, additional training may include connecting hose to hydrants, laying hose, coupling hose sections, and attaching nozzles.

Fire extinguishers: Training should be provided annually in the use of fire extinguishers for different classes of fire and hands-on training in their operation should be provided. For incipient fire fighting, employees should be instructed to (1) report the fire first (2) attempt to extinguish the fire or hold it in check until automatic sprinklers activate (where provided) or the public fire department arrives and (3) evacuate the area if the fire cannot be extinguished (6).

Demonstration fires: Where local laws permit, fires can be staged in safe outdoor locations to stimulate fires that may occur in a particular part of the facility. Small pans of flammable liquids on fire can be used as practice for possible fires caused by flammable liquid spills in a process area. Storage-type fires can be simulated by using cartons, crates, and combustible scrap stock piled as in a storage area. Allowing employees to practice on a small demonstration fire will give them confidence to take action against a real fire in their area.

Sprinkler valve and fire pump operation: Where sprinklers and fire pumps are provided, designated operators must be trained in the mechanism and operation of equipment. These people should be trained in the use of this equipment before an emergency occurs. This type of specialized training is also necessary for those assigned the responsibility for electrical and steam and gas piping control.

Isolation building areas: Knowledge of the location and operation of fire doors is necessary. Manual closing of these doors may be necessary for safe evacuation and to limit fire spread throughout the facility.

If a fire brigade is necessary, additional training will be needed. There are many excellent fire schools in the United States that can train fire brigades. Some of these schools are listed in the OSHA regulations, 29 CFR 1910, Subpart L—Fire Protection (5). The training should include classroom instruction, review of emergency action procedures, prefire planning, review of special hazards in the workplace, and practice in the use of self-contained breathing apparatus.

Fire brigades that are expected to perform emergency rescue and interior structural fire fighting should receive training in the proper techniques in rescue and fire suppression procedures. Training should include simulated fire situations including "wet drills" and, when feasible, extinguishment of actual mock fires. The training should be at least quarterly, but some drills or classroom training may need to be conducted monthly to maintain proficiency of fire brigade members.

Training Program in Writing

The training program should be in writing, and a manual or handbook is useful to describe the job assignments, schedules, facility hazards, equipment, and emergency action procedures (2). The following gives a brief outline of items that could be covered.

Organization
    Description of emergency assignments
    Assignment chart
    Schedule for drills, classroom meetings, inspections
    Use of first-aid, protective equipment, other equipment
    Fire prevention plan
Facility
    Plan of property, showing hydrants, valves, fire department connections, alarm panels, exits, etc. (see plan section in Chapter 8)
    Listing of specific hazards and how to handle emergency: cutting and welding, flammable liquids, storage, dip tanks, etc.
    Listing of "no smoking" areas
    Listing of fire protection rules
Equipment
    Sprinkler control valves
    Principles of operation of automatic suppression equipment

Alarm system
Portable fire extinguishers
Standpipe and hose systems
Emergency shutoffs for processes and piping
Emergency Action
    How emergencies are handled
    Emergency action plan
    Checking suppression systems and fire pumps
    Fire attack
    Coordination with public fire department and emergency personnel
    Salvage
    Restoring systems

EMERGENCY ACTION PLANS AND FIRE PREVENTION PLANS

Where emergency action plans and fire prevention plans are required and provided (see Chapters 3 and 4), the employees should be trained in the implementation of the plans. For the emergency action plan, the evacuation wardens should
be trained in the complete facility layout and the various evacuation routes. The
plan should also be reviewed by each employee covered by the plan to ensure that
they know what is expected of them in all emergency possibilities. This should be
done initially when the plan is developed, whenever an employees's responsibilities or designated actions under the plan change, and whenever the plan changes.

The fire prevention plan should be reviewed with all employees when the plan
is initially developed to ensure that they are aware of the types of fire hazards of
the materials and processes to which they might be exposed.

REFERENCES

1. Cote, Arthur E., ed., *Fire Protection Handbook*, 16th Edition, National Fire Pro-
    tection Association, Quincy, Mass., 1986.
2. *The Handbook of Property Conservation*, 3rd Edition, Factory Mutual Engineer-
    ing Corporation, Norwood, Mass., 1983.
3. Smith, L. C., "The Ingredients of an Organized Training Program," *National
    Safety News*, (no date).
4. Mager, Robert F., *Preparing Instructional Objectives*, 2nd Edition, David S. Lake
    Publishers, Belmont, Calif., 1984.
5. *General Industry Standards—29 CFR 1910*, Occupational Safety and Health Ad-
    ministration, Washington, D.C., 1988.
6. Linville, James L., ed., *Industrial Fire Hazards Handbook*, 2nd Edition, National
    Fire Protection Association, Quincy, Mass., 1984.

13
Preventive Maintenance in Fire Protection

The term *preventive maintenance* obviously is used very extensively throughout industry, but is very seldom used when referring to the fire protection field or fire protection equipment. Preventive maintenance can be described as those actions taken on a regularly scheduled basis that will ensure the proper performance of any particular piece of equipment. Although we in fire protection do not use the term preventive maintenance, what better term can be applied to actions taken to ensure the proper performance of fire protection systems and equipment?

It is generally recognized that the alternative to what can be considered a preventive maintenance program for fire protection is "breakdown" maintenance. This can manifest itself in the failure of protective systems at the time of ignition of combustibles, resulting in the possibility of large or excessive losses. Preventive maintenance programs can be instituted for three principal reasons: (1) to extend life of equipment, (2) to minimize breakdowns, and (3) to provide smoother, more reliable performance. To the manager, all these reasons (and consequently, a preventive maintenance program) will be of utmost importance to the overall loss control effort.

Fire protection equipment preventive maintenance should be a regular part of the overall preventive maintenance program at the facility. A good program should include the following basics (1, 2):

1. Documented organization
2. Written program, including equipment inventories, schedules, and actions

3. System of developing and keeping essential records
4. Controls to supervise the program for completeness, timeliness, and effectiveness

As with the fire loss control program, the full support and commitment of top management are essential. A policy statement should be issued that clearly defines the facility's attitude and position on preventive maintenance. This can be part of the overall loss control program statement as outlined in Chapter 3.

Along with this policy statement, a maintenance manual is recommended. This manual can be simply a compilation of the facility's policies and practices. The manual can cover subjects such as maintenance organization, planning and scheduling, work order systems, preventive maintenance, and training (1, 2).

There are two important elements in any preventive maintenance system for a fire protection system: (1) a comprehensive self-inspection program done on a regularly scheduled basis, and (2) a regularly scheduled testing program. Both of these items have been referred to previously and their importance highlighted, particularly under the section referring to the human element areas of fire protection in Chapter 3.

The self-inspection program will point out obvious problems with fire protection systems, which could include closed valves, a loss of power to a fire pump, obstructed sprinkler heads, the loss of heat in sprinklered areas, and many others. Part of this program is establishing an equipment list. This list should at least include the type of equipment, functional description, identification numbers, location code, service or technical manual location, and year of manufacture. The frequency of the self-inspection program will depend upon several factors, including the extent of fire protection provided, the values at risk, the hazards and processing involved, and others. While a nonsprinklered facility, such as a distribution warehouse, depending upon extinguishers for protection, might conceivably require only monthly self-inspections (assuming good housekeeping and other good loss control factors), a sprinklered facility or a facility having special protective system should have weekly self-inspections to ensure the integrity and reliability of the systems.

An important standard to ensure the proper maintenance of the single most important protection system is the *Recommended Practice for the Inspection, Testing and Maintenance of Sprinkler Systems*, NFPA 13A (3). As we have stated previously, the key to the proper maintenance of a sprinkler system, as with other protection systems, is the self-inspection program, and weekly inspections should be made along with a schedule of testing. The weekly inspection should include drain tests and, less frequently, the opening of the inspector's test connection for testing the lines. Drain tests consisting of opening the 2-in. angle valve at the sprinkler riser should be done only when weather conditions permit, not during subfreezing temperatures.

NFPA 13A not only provides detailed guidelines for care and maintenance of sprinkler systems, but also indicated the steps that should be taken to properly maintain water tanks, whether they are gravity, suction, or pressure tanks; to maintain fire pumps; and to check and test fire hydrants.

With reference to automatic sprinkler system maintenance, perhaps the single most important item is to ascertain that control valves are in a wide open position. Valves should be sealed or locked wide open even if central station or proprietary valve tamper switches are included.

Of major concern in sprinkler system maintenance are obstructions by stock, usually caused by high piling or by building partitions that interfere with the proper distribution of water discharged through sprinklers, and the severe problems that can be encountered with freezing of sprinkler equipment even in small or what may be considered unimportant areas. Other important points to consider are

Checking of sprinklers regularly to ascertain if heads are in good condition with no corrosive loading, paint deposits, or damage
Checking to ensure that hangers are not broken or loose
Checking of dry-pipe system air pressure and heat in the dry-pipe valve enclosure

With regard to the maintenance of tanks furnishing a water supply for fire protection, whether gravity, suction, or pressure tank, the single most important factor to consider is the possibility of freezing. With growing emphasis on energy conservation, many plants have sought to save on fuel and have considered possible areas where this might be accomplished. One of these areas has been the heating of water tanks, which heretofore generally have been heated to 42° F (5.6° C). This certainly would seem to be a legitimate area in which energy could be reduced. However, any energy reduction in this direction requires very close supervision in the form of inspection and/or low-temperature alarms. Certainly gravity tanks, particularly if temperatures are allowed to stabilize at a level of say, 35 to 37° F (1.7 to 2.8° C), require almost daily inspections to see that ice formations are not occurring.

Proper preventive maintenance of water tanks will also include anticorrosion actions. For the most part these will take the form of an examination of not only the exterior but also the interior of steel tanks without cathodic protection to determine the condition of the tanks and the need for recoating. Efforts to ensure that tanks are constantly filled with water can also be considered a preventive maintenance item.

An important example of how testing can disclose the need for corrective action on sprinkler systems is the operation of the inspector's test connection and the drain test. The clogging of the inspector's test connection and continued discoloration of the water from the drain test will indicate the possibility of a condition that could result in obstruction of portions of the systems and plugging of some heads during fire, which could lead to a failure of the system to properly control or extinguish a fire. The most frequent obstruction of sprinkler piping is from scale from a rusting condition or also occasionally other items such as stone, dirt, etc., resulting from improper flushing and cleaning of the systems. Pipe scale is particularly noted in dry-pipe systems that are alternately maintained wet and dry. This condition can be avoided by maintaining dry-pipe systems "dry" throughout the year.

Where there is a reason, as a result of testing, to believe that obstructions do exist, a thorough investigation should be made of the plant underground mains and the suction supply for the fire pump, if such equipment is provided, along with checking of the interior of tanks. Following that, the sprinkler system should be checked be investigation of branch lines for obstructions. Once it has been determined that obstructions exist, the systems should be properly flushed using one of the two accepted methods, either the hydraulic or hydropnuematic procedures. Simply stated, the hydraulic method is accomplished by flushing sprinkler piping with water through hose attached to the ends of the pipe, using velocities of sufficient magnitude to loosen and remove foreign materials. The hydropneumatic method uses a stream of water and compressed air to gain the same objective.

There is very little that can be done from a preventive maintenance standpoint for underground piping other than occasional flushing. Leaks will occur and corrosion will take place that the preventive maintenance efforts cannot prevent. These problems may have been caused at the time of installation, created by poor workmanship or a corrosion problem from unfavorable soil conditions and the selection of non-corrosion-resisting pipe. However, hydrants and hose are another matter, and the self-inspection program should include both at least on a monthly basis. Hydrant valve stems should be lubricated annually, and hydrants should be opened at least annually and preferably twice a year in the spring and fall. All rubber-lined hose should be hydrostatically tested annually, and couplings and nozzles should also examined for damage.

Special protective systems such as carbon dioxide, dry chemical, and halon systems should be part of the overall fire protection preventive maintenance program. Again, the preventive maintenance program starts with adequate testing of these systems, followed by complete servicing on an annual basis. Due to the specialized knowledge generally required for the proper servicing of such systems, it is suggested that service contracts be let to qualified firms for performance of these portions of the preventive maintenance program.

Annual servicing should include complete inspection and servicing for proper operation of all phases of the system, including the detection system, the alarms, actuation of the control valves, and selector valves, if any. In the case of carbon dioxide containers and those used for storage of dry chemical or halon, semiannual tests should be made for loss of gas by carefully determining any reduction in weight or pressure. Some of these functions can be delegated to plant personnel; however, this should be done only after the employees involved are thoroughly competent in the complete servicing of the function under their control.

Maintenance of fire extinguishers is certainly an important segment of any overall maintenance program for fire extinguishing equipment. This begins with a monthly inspection of the units with quick visual observation to detect any obvious defects and to ascertain that extinguishers are in their designated locations and then involves periodic servicing.

The more detailed maintenance of extinguishers, generally done on an annual basis, requires the services of competent, trained individuals in this field, whether available in-house or from a service organization. Proper servicing of all types of extinguishers is completely outlined in the *Standard for Portable Fire Extinguishers*, NFPA 10 (3) and includes check points in maintenance and corrective actions. Extinguisher recharging should be done after use, during annual servicing, or when inspection indicates this need. It should be noted that stored pressure-type extinguishers provided with pressure gauges do not require annual maintenance when inspected monthly.

Extiguishers should be hydrostatically tested at 5-to 12-year intervals, depending upon the type. Procedures set forth in NFPA 10 should be followed. It is important, not only in hydrostatic testing, but also in inspections and annual servicing, to record all of these activities.

Automatic fire detection and alarm systems should be on a maintenance and testing program. Automatic fire detectors should be tested on a periodic basis. For example, restorable spot-type heat detectors should be tested semiannually (one or more detectors per circuit) so that, over a five-year period, all of the detectors will have been tested. For nonrestorable spot-type heat detectors, after the 15th year of operation, two detectors per 100 detectors should be removed (replaced with new detectors) and sent to a testing lab for evaluation. This should be done every five years. All smoke detectors should be tested at least semiannually (3). Guidelines for tests and maintenance on alarm systems can be found in the *Standard on Automatic Fire Detectors*, NFPA 72E, and the *Guide for Testing Procedures for Local, Auxiliary, Remote Station and Proprietary Protective Signaling Systems*, NFPA 72H (3).

This chapter on the proper maintenance of fire protection equipment is not intended to be a complete handling of the subject, but is designed to remind those who have responsibility for this task of the necessity for properly maintaining fire protection equipment to ensure its proper performance at the time of an emergency. It is not complete because there are many other areas that should be included in the preventive maintenance program, including proper electrical maintenance. The *Recommended Practice for Electrical Equipment Maintenance*, NFPA 70B (3), states that its purpose is to reduce hazard to life and property that can result from failure or malfunction of industrial-type electrical systems and equipment. The standard should be adopted into the overall plant preventive maintenance program and duties should be assigned to qualified personnel in the maintenance or electrical department.

There are other areas that should be included in an overall fire protection preventive maintenance program. One such area is the maintenance of protection for steel such as concrete block, sprayed-on mineral fiber protection, etc., which give steelwork the desired degree of fire rating. Also included is the assurance that fire subdivisions are properly maintained by proper maintenance of fire doors and that all penetrations of fire walls and barriers are provided with an adequate level of protection. Another area that might be considered part of the fire protection preventive maintenance program is occupancies involving hazardous materials such as flammable liquids. Here the program could involve the maintenance of ventilating systems designed to keep concentrations of flammable liquid vapors well below the lower flammable limit and the proper maintenance of static grounding and bonding and also transfer systems, including pumps and flammable liquid-handling safety containers.

In the safeguarding of lives and property, great dependence is placed upon fire protection equipment and systems that have been provided. Their proper functioning when needed to provide that vital first line of defense against serious loss can be ensured only by a comprehensive self-inspection and preventive maintenance program.

REFERENCES

1. "Preventive Maintenance, A Guide for Planning Your Program," p8414, Factory Mutual Engineering Corporation, Norwood, Mass., 1985.
2. "Preventive Maintenance, Your Role in Planning Corporate Strategy," p8319, Factory Mutual Engineering Corporation, Norwood, Mass., 1984.
3. *National Fire Codes*, National Fire Protection Association, Quincy, Mass., 1989.

14
Regulatory Contacts

American industry constantly encounters governmental regulations affecting the conduct of their operations. While the emphasis of many of these regulations is on employee health and safety, fire protection and prevention codes and laws are also involved.

In business or industry today, in the process of starting a new operation from the ground up or in acquiring an operation, management must interface with all levels of government, including federal, state, and municipal. Each one of these groups has adopted codes or laws that require certain consideration in construction and design and in the operation or conduct of business with regard to the health and safety of employees.

BUILDING CODES

Building codes set minimum requirements for design and construction of buildings and structures, including structural design, fire protection, means of egress, light, sanitation, and interior finish. The primary objective of building codes is to provide for the safety of occupants within the building or structure.

In building codes, occupancies are classified with regard to fire severity; in addition, various construction categories are classified. In Chapter 8, the classification of buildings using the generally accepted NFPA approach was discussed. Building codes do not necessarily follow these classifications, but use their own designations, which are, however, rather closely aligned with the NFPA classifications.

Occupancies are classified according to fire severity and also the major type of occupancy encountered. A list of the various occupancy groups according to their life safety hazard is contained in Chapter 4, Life Safety Elements. Building code occupancy classifications are quite similar, classifying occupancies as office, institutional, assembly, and others.

Building codes also specify building areas and height limitations for the various classifications, along with construction details and materials; acceptable interior finishing flammability limits; the enclosure of vertical openings, which certainly is important both for evacuating people safely from a fire and for retarding the spread of fire; and exit requirements. Codes also indicate where automatic sprinkler protection or other protection is required and where they can be used as an alternative. These alternatives, also known as "trade-offs," can frequently represent a meaningful savings in design and construction, permitting the installation of automatic sprinklers by the savings realized. Trade-offs for automatic sprinklers, which are considered by fire protection engineers to provide a better protected building and greater life safety, are the lessening of fire resistance ratings for building structural members, the use of less costly building materials, the elimination of fire walls by permitting greater fire areas or unlimited fire areas, a possible increase in the building height, and an increase in travel distance to exits, which will permit the elimination of fire walls or other costly considerations in providing for exits in large, undivided area buildings.

Due to the complexities of modern building code development and to help communities, there are three model building codes in use throughout the United States. These model codes have stood up well over the years for the average community. These codes have been developed by associations of building officials and are published on a periodic basis. Each organization and its model code are discussed in the following paragraphs.

Uniform Building Code

The *Uniform Building Code* is published by the International Conference of Building Officials (ICBO), Whittier, California. This code was first published in 1927 and is principally used in the western United States. ICBO also publishes mechanical, plumbing, and fire prevention codes. Amended versions of the building code are published every three years (1).

Standard Building Code

The *Standard Building Code* is published by the Southern Building Code Congress International (SBCCI), Birmingham, Alabama. First published in 1945 as the *Southern Standard Building Code*, this code is used mainly in the southern states. SBCCI also publishes mechanical, plumbing, fire prevention, and gas codes. Amended versions are published every three years (2).

Basic/National Building Code

The *Basic/National Building Code* is published by the Building Officials and Code Administrators, International (BOCA), Country Club Hills, Illinois. The code was first published in 1950 as the *Basic Building Code*. In addition, BOCA acquired the use of the name "National Building Code" from the American Insurance Association, which published the *National Building Code* from 1905 to 1976. This code is used in the midwest and northeastern states and is reprinted every three years. Like the other code organizations, BOCA also publishes mechanical, plumbing, and fire prevention codes (3).

FEDERAL AGENCIES

Federal agencies have been active in fire protection for many years, in both a regulatory and nonregulatory role, providing for protection of employees, the public, and federal property. In recent years, federal regulatory authority for safety has greatly increased. Federal authority designed to protect the public at large, such as the Consumer Product Safety Act and, most recently, the Emergency Planning and Community Right-To-Know Act, or to protect major segments of society, such as the Occupational Safety and Health Act, which applies to employees in the workplace, are having a major impact. The majority of the federal agencies have headquarters in Washington, D.C., with most having 8 to 10 regional offices throughout the United States. Brief descriptions of some of the agencies follow.

Occupational Safety and Health Administration (OSHA), 3rd Street and Constitution Avenue, NW, Washington, D.C., 29210

The Occupational Safety and Health Administration administers the occupational safety and health standards that have been referred to in earlier chapters, most notably the *General Industry Standards—29 CFR 1910*. In the fire protection area, these standards cover means of egress, emergency action plans, fire prevention plans, hazardous materials, fire brigades, portable fire extinguishers, standpipe and hose systems, automatic sprinkler systems, fixed extinguishing systems, and fire detection and employee alarm systems.

Federal Emergency Management Agency (FEMA), 500 C Street, SW, Washington, D.C., 20472

United States Fire Administration (USFA) This agency is involved with fire policy and coordination, fire-fighter health and safety, fire data analysis, and fire prevention and arson control. USFA developed the National Fire Incident Reporting System, which is being used to collect data in over 40 states. The Public Education Office of USFA has provided extensive assistance to help communities and states develop comprehensive public fire education programs.

National Fire Academy (NFA)   The National Fire Academy is involved with training in fire prevention management, fire prevention and loss control, and fire service technology.

Environmental Protection Agency (EPA), 401 M Street, SW, Washington, D.C., 20460

The Environmental Protection Agency has broad regulatory, enforcement, and research authority relative to reducing air and water pollution. Some aspects may have an impact on fire safety. For example, the use of some fire extinguishing agents might be limited, such as halon systems. The Community Right-To-Know regulations have an impact due to the physical fire hazards of chemicals being manufactured, used, or stored by industries.

Department of Transportation (DOT), 400 7th Street, SW, Washington, D.C., 20590

Office of Hazardous Material Transportation   This office enforces regulations for safe transportation of hazardous materials and the labeling and identification of hazardous materials. In addition, the Transportation Safety Institute provides training on improved safety and security management of transportation functions.

Office of Pipeline Safety   This office enforces the safety standards for the transportation of gas and hazardous liquids by pipeline.

National Institute for Occupational Safety and Health (NIOSH), 944 Chestnut Ridge Road, Morgantown, WV 26505

The National Institute for Occupational Safety and Health is involved in research in the occupational safety and health field and recommends standard criteria to support OSHA's activities. Fire protection plays a major role in its safety research, including recent research in areas such as explosions and fires in grain-handling facilities, oil and gas well drilling operations, liquefied petroleum gas operations, flammable liquids, and liquefied natural gas facilities.

Consumer Product Safety Commission (CPSC), 1111 18th Street, NW, Washington, D.C., 20207

The Consumer Product Safety Commission has broad regulatory authority over safety and health aspects of products sold to consumers. This agency is also involved with flammability standards for fabrics.

National Academy of Sciences (NAS), 2101 Constitution Avenue, NW, Washington, D.C., 20418

This quasi-governmental organization, operating under charters from Congress, provides a focal point for research in several aspects of the fire problem. The Committee on Evaluation of Industrial Hazards has stimulated research, conducted seminars, and published reports in fire protection areas, such as classification of hazardous materials with reference to the *National Electrical Code*, hazardous locations, grain elevator and mill explosions, pneumatic dust control, and investigation of grain dust explosions.

REFERENCES

1. ICBO, *Uniform Building Code*, 1985 edition, International Conference of Building Officials, Whittier, Calif., 1985.
2. SBICC, *Standard Building Code*, 1985 Edition, Southern Building Code Congress International, Inc., Birmingham, Ala., 1985.
3. BOCA, *Basic/National Building Code*, 1984 Edition, Building Officials and Code Administrators, International, Country Club Hills, Ill., 1984.

Appendix

Conversion Table

English to Metric (SI) Units Commonly Encountered in Fire Protection

Category	Given	Multiply by	To obtain
Length	Inches	2.54	Centimeters
	Feet	0.3048	Meters
Area	Square feet	0.0929	Square meters
Volume	Cubic feet	0.0283	Cubic meters
	Gallon (U.S.)	0.0038	Cubic meters
	Gallon (U.S.)	3.785	Liters
Mass	Pounds	0.4536	Kilograms
Pressure	Pounds per square inch	0.0689	Bars
	Pounds per square inch	6894.76	Pascals
	Pounds per square foot	47.8803	Pascals
	Pounds per square inch	70.307	Grams per square centimeter
Velocity	Feet per second	0.3048	Meters per second
	Feet per minute	0.3048	Meters per minute
Water flow	Gallons (U.S.) per minute	3.785	Liters per minute
Hydraulic density	Gallons per minute per square foot	40.746	Liters per minute per square meter

Index